BRYOZOANS

Biological Sciences

Editor
PROFESSOR A. J. CAIN
MA, D.PHIL
Professor of Zoology
at the University of Liverpool

BRYOZOANS

J. S. Ryland
MA, PH.D
Senior Lecturer in the Department of Zoology,
University College of Swansea

HUTCHINSON UNIVERSITY LIBRARY
LONDON

HUTCHINSON & CO (*Publishers*) LTD
178–202 Great Portland Street, London W1

London Melbourne Sydney
Auckland Johannesburg Cape Town
and agencies throughout the world

First published 1970

This book has been set in Times type, printed in Great Britain on smooth wove paper by Anchor Press, and bound by Wm. Brendon, both of Tiptree, Essex

ISBN 0 09 103870 7 (cased)
0 09 103871 5 (paper)

CONTENTS

FIGURES

PREFACE

Bryozoans are common and fascinating aquatic animals, which are also found abundantly as fossils. The phylum Bryozoa (or Polyzoa) is of medium size, with approaching 4000 extant species and perhaps four times as many preserved as fossils. Such numbers are not greatly different from those relating to the Echinodermata, for example, so why are bryozoans less well known than the starfishes and their allies? The answer lies mainly in their smaller size. Bryozoans, although present in fresh and brackish waters, and especially numerous in the sea, must be looked for, and their beauty revealed by the microscope. Anyone who starts looking at bryozoans will continue to do so, for their biology is full of interest and unsolved mysteries.

Growing interest in the Bryozoa is reflected both by the recent steep rise in the number of papers about them published each year, and by the formation of the *International Bryozoology Association.* The Association held a very successful 1st Conference in Milan during the summer of 1968, the *Proceedings* of which provide a valuable insight to current trends in thought and research on the Bryozoa.

Dr Libbie Hyman, in Volume 5 of *The Invertebrates*, admirably summarized our zoological knowledge of bryozoans as it stood a decade ago. Yet the student may well be deterred from reading her account by the amount of morphological and other detail she describes. Professor Paul Brien has also written a comprehensive and valuable zoological review, especially of the freshwater class Phylactolaemata, in Tome 5 (2) of the *Traité de Zoologie.* Below treatise level there exists no account that can be recommended to the advanced student or teacher.

For palaeontologists the situation is worse. The accounts in the standard textbooks, while conveying the basic facts about the geological history of the Bryozoa, are in other ways somewhat unsatisfactory. Moreover, great strides have been made recently towards a better understanding of the structure and mode of life of bryozoans belonging to the extinct orders.

My task was to present, for the first time, an account of the Bryozoa which was broad-based in conception but uncluttered by detail: to bring together as a unified narrative the work of biologists and palaeontologists, and to give prominence to the fundamental characteristics and evolutionary trends which distinguish the Bryozoa. This book is the result. I do not presume to guide the rapidly increasing number of practising bryozoologists in their own speciality, but I hope to introduce marine biologists, geologists and zoologists to a group of animals which deserves more attention, and to provide a text which fulfils the needs of university students studying these three subjects.

I have been encouraged and greatly assisted by the interest shown by my colleagues, by discussions with many of them at the Milan conference, and by their willingness to read and suggest improvements to sections of the draft. Miss Pat Cook, Dr D. E. Owen, Dr E. Naylor, Dr June Ross and Dr R. Tavener-Smith have assisted me in this way; their guidance on the extinct orders has been particularly valuable. Dr R. S. Boardman, Dr A. H. Cheetham, Dr D. Eggleston, Mr D. P. Gordon, Dr Anna Hastings, Mr L. J. Pitt, Mr A. R. D. Stebbing and Dr R. Tavener-Smith kindly placed manuscript notes, theses or then unpublished papers at my disposal. I am especially indebted to Dr Hastings, who not only discussed parts of the book in early draft but later read and commented on the entire typescript, and to Professor A. J. Cain who also read the typescript and suggested a number of improvements. It is my pleasure to thank all these people, together with others not specifically mentioned, who have contributed to making this book; at the same time emphasizing that all the contents, together with any omissions, are solely my responsibility.

I

INTRODUCING THE BRYOZOA

At a casual glance a bryozoan colony might be mistaken for a hydroid or coral, or even for a piece of seaweed. But examine it under water with a dissecting microscope. Observe the hesitant emergence of the bells of tentacles for feeding, and the changing patterns of iridescence as the beating cilia scatter and refract the light. The presence of cilia on the tentacles is sufficient in itself to distinguish a bryozoon from a hydroid. Each bryozoon is a compound animal made up of hundreds of zooids. In some species a calcified skeleton hides the inner structures, but in *Bowerbankia* or *Membranipora*, for example, much of the internal anatomy can be observed. Here is no hydroid-like simplicity, but a much more elaborate animal. Where are these bryozoa found, what is their structure, and how do they live?

Bryozoa are sessile organisms. A few inhabit freshwater lakes and rivers, but most are marine. Anyone who has collected animals on a rocky shore in western Europe will be familiar with *Membranipora*, which so often covers the fronds of kelps with its lacy white colonies; and they may have noticed the less conspicuous colonies of *Electra* and *Flustrellidra* (Fig.1A) on the mid-shore fucoids: but how many realize that careful searching might reveal twenty-five or more different species?

Yet most bryozoa are never found on the shore, for they occur much more prolifically below tidemarks. As many as thirty species have been recorded on a single *Pinna* shell, and up to ninety in one dredge haul.[68]* The free diving technique offers unrivalled opportunity for the ecological study of bryozoa for, while they extend into the ocean depths, their greatest abundance and diversity occur in the

* Superior figures in the text refer to references cited at the end of the book.

relatively shallow waters of the illuminated zone. Here are large tufts of the slender, branched *Cellaria*, of *Bugula*, and of the citron-smelling *Flustra* so often stranded along the high tideline (Fig. 1B,C,F,G). Here too are the rigid folded layers of *Pentapora* in clumps which may reach several feet in circumference, the fleshy pale brown fingers of *Alcyonidium* (Fig. 1E,H), and a host of others.

Alcyonidium is so common in parts of the North Sea that when a trawler's net is emptied, the bryozoans alone may form a pile several feet across. Repeated handling of this *Alcyonidium* can induce severe allergic dermatitis with a painful rash and large weeping blisters—an unpleasant and unexpected hazard for fishermen.[23]

Sedentary animals, bryozoa among them, soon colonize any free surface in the sea. When that surface is a ship's hull, the settlement is termed 'fouling'. As it results in loss of efficiency, fouling constitutes a serious economic problem. Over fifty species of bryozoa have been recorded on ships' hulls. The use of copper paint is a palliative, effective for a few months only, and certain bryozoa are among the most copper-resistant fouling organisms known. Their rapidly spreading colonies provide a safe substratum for the subsequent attachment of more sensitive forms.

Being filter-feeders, bryozoans do not need light to live, only food-bearing water. Tufted species such as *Bugula* and *Zoobotryon* may settle and prosper in the intake pipes of ships' and power stations' cooling systems, causing serious nuisance by restricting the flow of water; while freshwater species can block water supply pipes, and regularly did so prior to the introduction of sand filtration. One 60 cm diameter pipe in Belgium contained a lining of bryozoa 15 cm thick; while in Manchester, at the beginning of the present century, 700 tons of bryozoa were removed at one time from the mains.[142]

The calcified skeletons of bryozoa are extremely durable, and there is a rich record of fossil forms from the early Palaeozoic onwards.

Fig. 1 Colony form in some gymnolaemates

A *Flustrellidra* (formerly *Flustrella*) incrusting *Fucus* (×3/8)
B *Bugula* (×1/2)
C *Cellaria* (×1)
D *Cupuladria* (in section and surface view) (×2)
E *Alcyonidium* (×1/4)
F *Bugula* (species with spirally disposed branches) (×1/2)
G *Flustra* (×1/2)
H *Pentapora* (formerly *Lepralia*) (×1/4)
I *Sertella* (formerly *Retepora*) attached to frond of *Posidonia* (×1)
J *Myriapora* (×1)

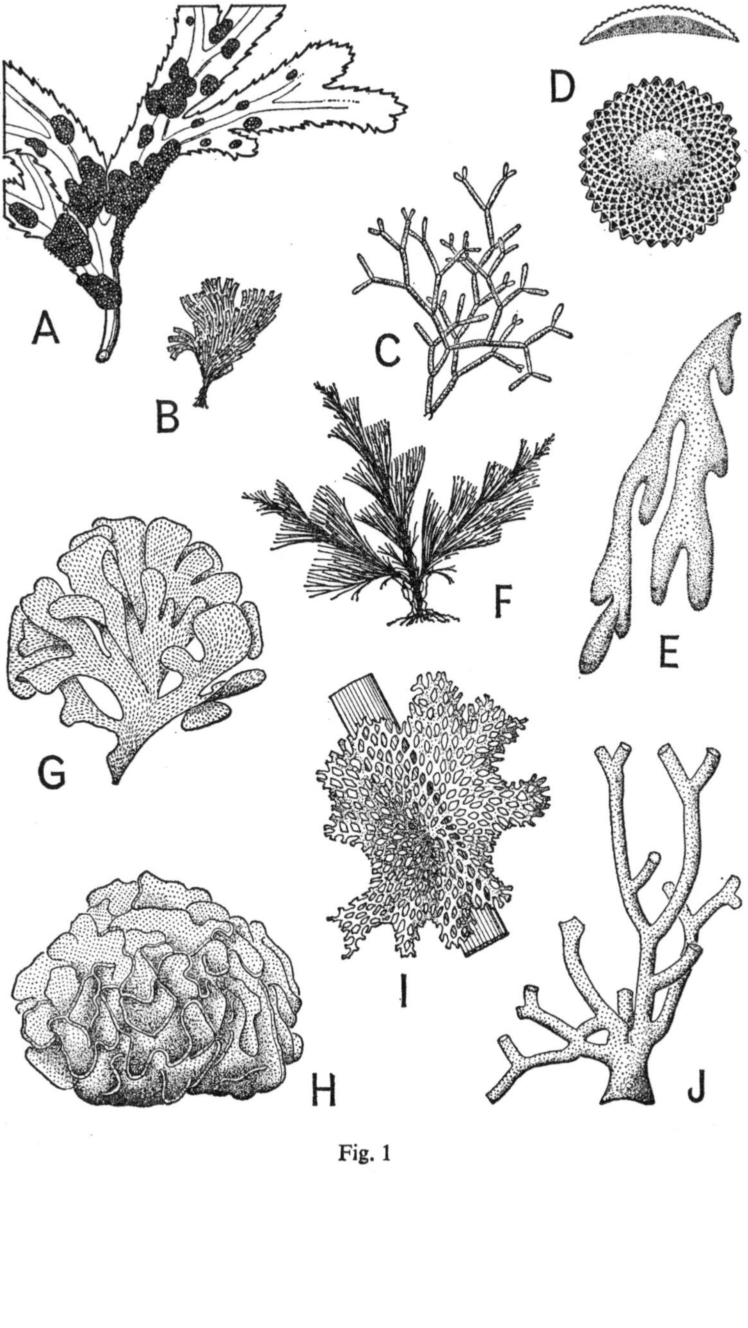

Fig. 1

Many geological formations are well characterized by their bryozoa, and the stratigraphic value of these and their application to economic problems in Geology have perhaps not yet been fully exploited, despite a voluminous literature on the subject. Covering the period of their long history, over 1200 genera have been described, and there are about 4000 species living today.[3]

HISTORICAL OUTLINE

John Ellis and the animal nature of corallines

Study of the internal anatomy of a bryozoon requires a microscope. It is not surprising that the early naturalists were unable to distinguish bryozoa from coelenterates, and indeed classified both, which were together known as corallines, as members of the plant kingdom. Ferrante Imperato, a Naples apothecary, first suspected that some corallines were partly animal in nature; but his *Historia Naturale* of 1599 seems to have had little influence on subsequent thinking.

In 1727 Réaumur presented to the Academy of Sciences in Paris the view of J. A. Peyssonnel, a Marseilles physician, that the corallines belonged to the animal kingdom. This opinion, being opposed to the views of the then eminent Count Marsigli who, having been the first to observe expanded polyps, claimed to have seen corals in flower, was greeted with derision; and Réaumur himself hastened to disown Peyssonnel's findings. Later, when Réaumur became aware of Abraham Trembley's discovery in 1741 of the first freshwater bryozoon, he changed his mind and became Peyssonnel's ardent supporter. His friend Bernard de Jussieu, who invented the term polyp, examined many forms, including *Flustra*, and presented a succinct memoir to the Paris Academy. This body remained unimpressed. Peyssonnel's views were also communicated to the Royal Society in London, where they met with just as much opposition.

The truth, however, was shortly to achieve proper recognition with the publication in 1755 of one of the great zoological books of the eighteenth century, the *Natural History of the Corallines*, by John Ellis.[12] Ellis, a London merchant, was by inclination a collector of plants, including seaweeds and corallines. It was his study of the latter with a microscope that revealed their animal nature to him. His meticulous researches, illustrated by admirably executed engravings, carried conviction where the unsupported statements of earlier naturalists had encountered scepticism. The excellence of Ellis' work was such that, in the 10th edition of *Systema Naturae* (1758), Linnaeus was able to base almost all his species of zoophyte solely on the plates and descriptions in the *Natural History of the Corallines*.[15]

The name ZOOPHYTA was invented by Linnaeus. Though greatly impressed by the conclusions of Ellis, he nevertheless suggested that the corallines in reality constituted a group intermediate between plants and animals. Such was Linnaeus' authority that the expression zoophyte remained in common use for over a century, although the organisms themselves were soon universally accepted as animals.

Early nineteenth-century classifications of the Zoophyta were based on the appearance of the colony, and many of the divisions established were a complete jumble of what are now recognized as hydroids and bryozoa. Then in 1820 de Blainville discovered that polyps of certain zoophytes possessed both mouth and anus, and proposed that these organisms should be placed on a higher level than the true polyps. A little later Grant observed that some zoophyte polyps had ciliated tentacles and a recurved alimentary canal.

J. V. Thompson's discovery of the Polyzoa

At about this time, J. Vaughan Thompson, already well known for his researches on the Crustacea, was studying zoophytes on the southern shores of Ireland. He discovered independently, in about 1820, that there were two anatomical forms of polyp, and published in 1830—unaware of the observations of Grant and others—a work headed *On Polyzoa, a new animal discovered as an inhabitant of some Zoophites* (recently reissued in facsimile).[25] The title reflects the then current view that the polyps were animals which lived in a calcareous, horny or fleshy matrix. Thompson's work shows that he really understood the structure of his POLYZOA; he was the first to propose a new name, and his intention was evidently to create a new animal class, placed in a radically different position in the classification from the Zoophyta.

In 1831, one year after the publication of Thompson's researches, and presumably without knowledge of them, G. C. Ehrenberg in Germany proposed the new names ANTHOZOA, for zoophytes with a single gut opening, and BRYOZOA, for those possessing both mouth and anus. Rather confusingly both of the names Polyzoa and Bryozoa remained in use, preference following a geographical pattern: a majority of zoologists in Britain and the Commonwealth countries have preferred Polyzoa, while those in America and on the mainland of Europe have used Bryozoa.

In the mid-nineteenth century the phylum as then understood included animals of two distinct types of organization, and these were distinguished by Nitsche in 1869 as BRYOZOA ECTOPROCTA and BRYOZOA ENTOPROCTA, according to whether the anus opened outside or inside the circlet of tentacles. The differences between the two were duly recognized as fundamental and, as long ago as 1888, Hatschek raised the ENTOPROCTA, as they were more simply called, to the level of phylum. Few today doubt the correctness of this opinion, but full recognition of the lack of close relationship between the Bryozoa and the Entoprocta has come about very slowly.

Hyman[17] suggested that by substituting the class name Ectoprocta as the name of the phylum, the existing controversy over the merits of the names Polyzoa and Bryozoa could be terminated. The issue is hardly an important one, and several other animal groups have alternative names. Moreover, substitution of the older names Polyzoa or Bryozoa by ECTOPROCTA, as Hyman proposed, violates widely accepted principles of zoological nomenclature[20] (though not the mandatory rules, which do not cover the naming of higher taxa). There is also the disadvantage that such closely similar words as Ectoprocta and Entoprocta can easily be confused, whether written, read or spoken. The original names are therefore to be preferred. Although Polyzoa undoubtedly has priority, it seems on balance desirable to follow majority usage and adopt Ehrenberg's term Bryozoa.

The need for subdivision of the true Bryozoa had been recognized in 1837, when Gervais separated the freshwater species from the marine as LOPHOPODA and STELMATOPODA. However, the class names generally employed are PHYLACTOLAEMATA (freshwater) and GYMNOLAEMATA (marine) proposed by Allman in his *Monograph of the fresh-water Polyzoa*.[137] Shortly before this, George Busk had introduced the ordinal names CYCLOSTOMATA, CTENOSTOMATA and CHEILOSTOMATA that form the basis of the classification of all extant marine bryozoans.

Bowerbankia—A BASIC BRYOZOON

A bryozoan *colony* is made up of large numbers of intercommunicating *zooids* which, in the simplest cases, nearly all resemble each other. Bryozoa may have arisen from a worm-like ancestor (see p. 24): at least, modern species with cylindrical zooids display bryozoan anatomy at its simplest. *Bowerbankia*, a genus of world-wide distribution, found in Europe on the fucoids of sheltered shores, has zooids of this form (Fig.2, p. 20). Moreover, the relative transparency of the body wall makes it suitable for laboratory study.

The zooid

The colony of *Bowerbankia* consists of a loose tangle of *stolons* from which clusters of zooids arise at intervals (Fig.2A). Each zooid, which is delimited by the smooth cuticular covering of the body wall, rises to a height of about 1 mm from its point of origin on the stolon. If observed alive in a bowl of water, many zooids will display at their free end a series of eight or ten slender, ciliated, outwards-curving *tentacles* (Fig.2A,B), which arise from an annular *lophophore* (*lophos*, crest or comb+*phoreus*, bearer) surrounding the mouth. The cilia create the water currents which carry diatoms and other particles towards the mouth. The *alimentary canal* makes a deep loop inside the coelomic cavity, with pharynx and oesophagus linking the mouth with the stomach, and the intestine and rectum rising to the anus which is situated near the mouth but outside the lophophore. Just below the lophophore, between the mouth and the anus, is the *nerve ganglion* (Fig.2B): this side is considered to be dorsal.

Surrounding the pharynx and terminal part of the rectum, and thus joining the lophophore to the lower, protective part of the zooid wall, is a cuticle-free region of the body known as the *tentacle sheath*, morphologically an introvert. If the water in the bowl be disturbed, the zooids will withdraw their tentacles: as these shoot back into the body cavity, the introvert rolls inwards like a cuff being pulled down inside its sleeve. When contraction is complete, the introvert completely ensheaths the tentacles and its name is explained (Fig.2F). Looked at now there are no structures outside the cuticularized zooid wall, which is broken only by the opening, termed the *orifice*, at its distal end.

A structure present only in *Bowerbankia* and allied forms is the *collar*, a membrane which encircles the lower part of the everted tentacle sheath. When the tentacles are withdrawn, the collar folds away in a series of pleats, effectively blocking the *atrium* or space just inside the orifice. Below the collar is the *atrial sphincter muscle*, by which the opening to the exterior is tightly closed (Fig.2B,C,F).

The body wall consists of two cell layers, an inner diffuse peritoneum and an outer compact epidermis immediately below the cuticle. In many bryozoa, though not in *Bowerbankia*, the epidermis secretes a calcareous deposit within the inner part of the cuticle, thereby greatly strengthening the zooid wall.

It is perhaps not surprising that some of the nineteenth-century naturalists, watching protrusion and retraction of the lophophore and tentacles, imagined that they observed the emergence of a feeding zooid which inhabited a protective house or cell. The former, 'the retractile portion of the Polyzoa', Allman termed the *polypide* (*polypous*, a many-footed—i.e. tentaculate—animal+*eidos*, resemblance) to emphasize its distinction from a hydroid polyp. The cell was named the *zooecium* (*zoon*, animal+*oikion*, dim. of *oikos*, house), according to its supposed role as a shelter. A little later, as an alternative, the zooid walls were called the *cystid* (*kystis*, cell or sac), a name which has the advantage of being descriptive while not implying a function. As far as possible the term zooecium, with its obsolete connotation, should be avoided in biological writing, although it is still widely used—and often not even in its proper sense—in the literature describing bryozoa.[23] If used at all, it is best restricted in meaning to the skeleton alone, for it is convenient to have a name for that part of the zooid which persists in dead and fossilized bryozoa.

Although we no longer regard the polypide as a separate entity dwelling in a protective case, there is some advantage in retaining the old-established term. The polypide comprises tentacles, tentacle sheath, alimentary canal, associated musculature and nerve ganglion. It is transient in the life of the zooid, sometimes degenerating and being replaced by a new one derived from the living layers of the body wall. Replacement of tissues occurs in all animals, and in bryozoa there must be advantages in this being an abrupt process. The cystid is more permanent, and the primordia which give rise to successive polypides and to gametes differentiate from it.

The body space of the zooid is considered to be a true coelom. A small ring-shaped lophophoral cavity, with extensions into the tentacles, is separated from the main coelom by a partition which is perforated by a single opening just by the ganglion. The U-formed alimentary canal (Fig.2B,C), which constitutes the major part of the polypide, consists of ciliated pharynx, oesophagus, tripartite stomach forming the base of the loop, intestine and rectum. In *Bowerbankia* there is also a gizzard, but this is a special feature of the genus rather than a typical character of bryozoa. Closely associated with the polypide is the *retractor muscle*, which arises near the proximal end of the cystid and divides into two bands so that its striated fibres can insert in a ring around the base of the lophophore (Fig.2B–E). It is largely the contraction of this muscle that withdraws the tentacles into the cystid in the manner mentioned earlier. Other muscles are associated with the cystid, and their role in the movements of the polypide will be discussed presently.

Descending from the stomach is a distinct cord of mesenchymatous tissue, known as the *funiculus* (*funiculus*, dim. of *funis*, rope), which passes through the base of the zooid and joins a similar strand running along the stolon (Fig.2B,C). The zooids, then, are not isolated individuals, but are interconnected throughout the colony by the funiculus.

There are no special respiratory organs, and no circulatory or excretory systems.

Zooids are hermaphrodite in *Bowerbankia*, as in many other bryozoa, with ovary and testis forming from the peritoneum of the cystid (Fig.2F,G). As the egg matures, the polypide starts to degenerate. The large yolky egg passes from the coelom through a minute *supraneural pore* at the base of the two most dorsal tentacles (Fig.2B), and undergoes development in the atrium[31] (Fig.2G). The yolk is brightly pigmented with yellow or red carotenoids,[71] and developing embryos are conspicuous objects in the colonies during late summer and autumn. After liberation and a short free life, the larva will settle, metamorphose, and give rise to a new colony.

The mechanism of tentacular protrusion

The cuticularized walls are flexible but inelastic, and contribute to the skeletal system of the zooid. The principal muscles are all either associated throughout their length with the walls, or have their origin on them. Thus, in the proximal part of the zooid, there are two vertical series of *transverse parietal muscles* (Fig.2B), small bands of smooth fibres adherent to the cystid. Their contraction pulls in the walls, thereby reducing the coelomic volume and raising the internal hydrostatic pressure. This pressure increase, with the concomitant relaxation of antagonistic muscles, everts the lophophore.

The transverse muscles are opposed by two sets of longitudinal muscles: the retractor, mentioned earlier, and the *longitudinal parietals*, which originate on the cuticularized wall at the distal end of the series of transverse muscles. The longitudinal parietals form two sets, each in four fascicles (Fig.2B,C,D,F,H): the smaller and deeper muscles go to the sphincter, and the broader muscles to the atrium.

Pressure in the coelom of the fully retracted zooid (Fig.2F) is probably already high enough to initiate eversion of the atrium by recoil as soon as the retractor starts to relax. As contraction of the transverse parietals raises the pressure still further, the longitudinal parietals relax. With the retractor now only partially contracted, the atrial wall everts (Fig.2C), and further relaxation of the retractor permits the tentacles to rise through the collar until they are wholly protruded (Fig.2B). They are then spread by the action of their intrinsic musculature (Fig.6C, p. 45). The process elegantly illustrates skeletal function by the transmission of pressures. Its success depends on the flexible but unextensible properties of the cuticle, the antagonistic muscle systems, and the mobility of the tentacle sheath which permits displacement of the polypide.

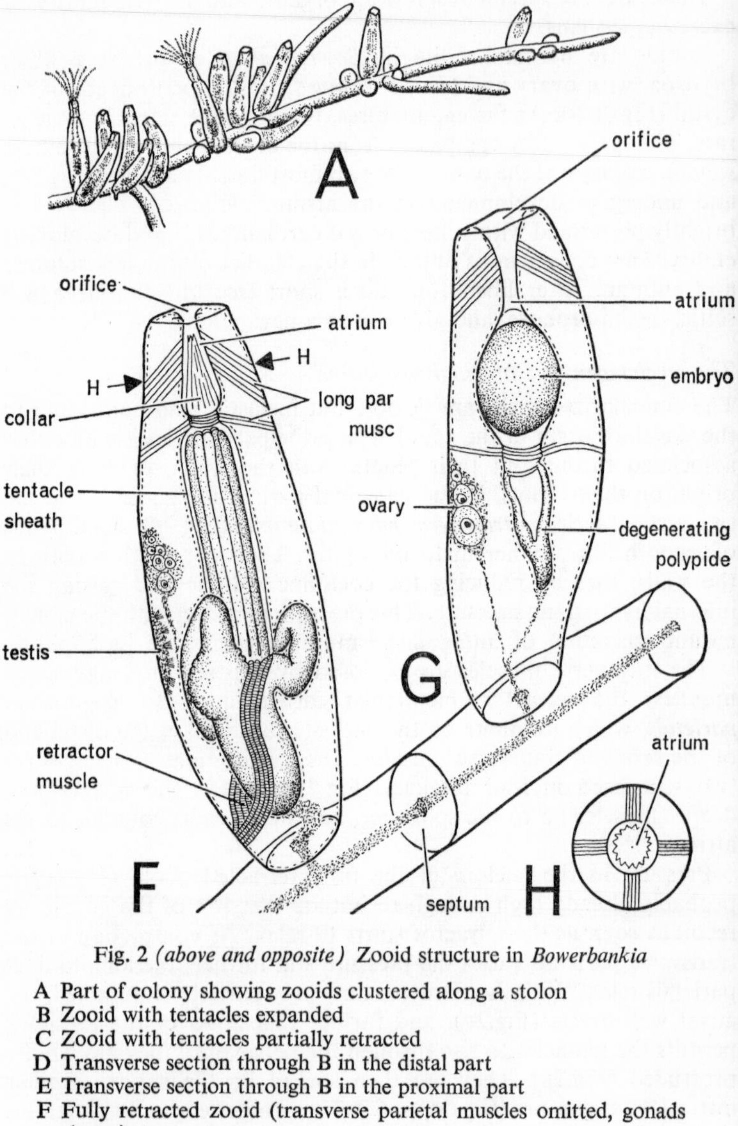

Fig. 2 *(above and opposite)* Zooid structure in *Bowerbankia*

A Part of colony showing zooids clustered along a stolon
B Zooid with tentacles expanded
C Zooid with tentacles partially retracted
D Transverse section through B in the distal part
E Transverse section through B in the proximal part
F Fully retracted zooid (transverse parietal muscles omitted, gonads shown)
G Zooid brooding embryo
H Transverse section through F in the region of the atrium

(Partly after Brien[7] and Braem[31])

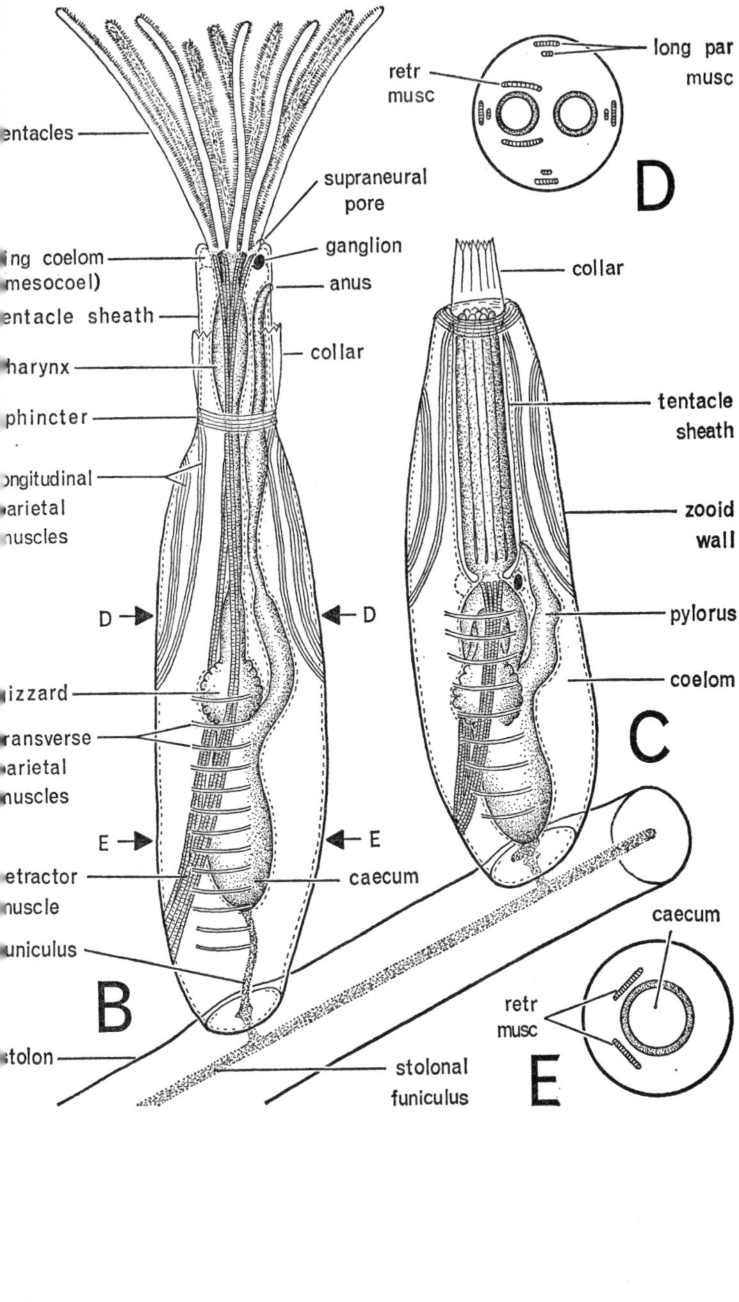

entacles
supraneural
pore
ing coelom
mesocoel)
ganglion
anus
entacle sheath
collar
harynx
phincter
ongitudinal
arietal
nuscles
D
D
izzard
ransverse
arietal
nuscles
E
E
etractor
nuscle
caecum
uniculus
B
stolon
stolonal
funiculus
long par
musc
retr
musc
D
collar
tentacle
sheath
zooid
wall
pylorus
coelom
C
caecum
retr
musc
E

Withdrawal of the tentacles is achieved by the extremely rapid contraction of the retractor, with the transverse parietal muscles relaxed. In the later stages retraction is accompanied by the sequential contraction of the two sets of longitudinal parietals. Finally, with the tentacles withdrawn, the circular muscles of the sphincter close off the polypide from the exterior (Fig.2F).

THE BRYOZOAN COLONY

Although the zooids of different bryozoa vary in shape and in the details of their structure, they can always be related to the same basic plan. With the colonies, on the contrary, this is far from true. Variations in form and texture seem almost infinite, and are related to fundamental differences in the pattern of budding and resultant growth, and to certain characteristics pertaining to the different orders of bryozoa.

Zooids sometimes grow in series or chains adhering to the substratum (*Aetea*, *Hippothoa*), or rising from it (*Scruparia*), or disposed along creeping or free stolons (*Bowerbankia*, *Zoobotryon*). Most commonly the colonies form flat incrustations of contiguous zooids placed in a single layer, and cover aquatic plants, stones, shells, and other substrata. They may make small roundish spots (*Escharella*, *Schizoporella*), lobed expansions (*Tubulipora*), or stellate patches (*Electra*); they may clothe the stems of algae (*Flustrellidra*, *Alcyonidium*) or cover great areas of frond (*Membranipora*). Less often the zooids may be jumbled together forming a mound rather than a flat crust.

A colony is normally anchored to the substratum, but may grow upwards to form variously lobed foliaceous or arborescent growths. The zooids may be arranged in a single layer (*Carbasea*, *Bugula*), in two layers adhering back to back (*Flustra*, *Pentapora*), or be disposed radially in a more or less cylindrical branch (*Alcyonidium*, *Cellaria*, *Myriapora*, *Porella*). The appearance depends much on the pattern of branching: short and broad (*Flustra*), narrow and frequently anastomosing (*Sertella*), strap-shaped in bushy tufts or disposed in a spiral (*Bugula*), or dividing dichotomously into short, jointed lengths (*Cellaria*).

Texture has an important effect on appearance. Some colonies form flaccid tufts (*Bowerbankia*) or are firmly gelatinous (*Alcyonidium*); others are tough but flexible (*Flustra*) or completely rigid from heavy calcification (*Myriapora*, *Pentapora*, *Sertella*).

Thus colonies (some of which are illustrated in Fig. 1) are almost bewildering in the variety of their formations and the catalogue of types could be continued. The examples illustrate that diversity of form, and therefore of adaptation to the environment, which is one of the attractive features of bryozoa. But whatever the shape, size, texture or colour of the colonies, their true identity will always be revealed by the structure of the zooids.

DEFINITION

Bryozoa are sedentary, aquatic, colony-forming coelomates. Each colony arises by asexual budding from a primary zooid or *ancestrula* formed by the metamorphosis of a sexually produced larva, or from a resting bud (*statoblast*) in the freshwater class Phylactolaemata; the zooids normally remain in communication throughout the colony. Each zooid has a circular or crescentic lophophore bearing a series of slender, ciliated, post-oral tentacles. The anterior part of the body forms an introvert, within which the lophophore and tentacles can be withdrawn. In the Phylactolaemata the mouth is covered by a projecting flap (*epistome*). The alimentary canal is deeply looped, so that the anus opens near the mouth, but just outside the lophophore. In most, the nervous system is represented mainly by a small ganglion between the mouth and the anus. Excretory organs are absent. There is no respiratory or circulatory system. Colonies, but not all zooids, are hermaphrodite; simple, ductless gonads, derived from the peritoneum, shed gametes into the coelom, which opens to the exterior through coelomopores. Each zooid secretes a rigid or gelatinous wall, so providing support for the colony. The class Gymnolaemata is well characterized by highly developed polymorphism of zooids. Most bryozoa live in the sea; a few inhabit brackish and fresh waters.

RELATIONSHIPS WITH OTHER PHYLA

Three groups of coelomates—BRYOZOA, PHORONIDA and BRACHIOPODA—resemble each other in the possession of a lophophore. Hyman[17] defines this structure as 'a tentaculated extension of the mesosome that embraces the mouth but not the anus, and has a coelomic lumen'. The lophophore is basically crescentic, as in *Phoronis*, primitive brachiopods and the Phylactolaemata: in other living bryozoa, where the adult lophophore is circular, the horseshoe form may be observed during development. PTEROBRANCHIA, POGONOPHORA and ENTOPROCTA also possess crowns of tentacles, but in none of these does the structure correspond to that of a lophophore.

Presence of a lophophore need not by itself indicate affinity between bryozoa, brachiopods and *Phoronis*; but there are several other features in common. In all three the body is considered to be basically tripartite, divided into *protosome*, *mesosome* and *metasome* as in the DEUTEROSTOMIA,* although the divisions are much less clear than in that group of phyla. In bryozoa, the *epistome* of phylactolaemates is supposed to represent the protosome. The lophophore develops from the mesosome as in brachiopods and *Phoronis*, and its cavity is the *mesocoel*. The rest of the body corresponds to the metasome, and the main coelom is a *metacoel*. In bryozoa the septum between mesocoel and metacoel is well developed.

The alimentary canal in all three groups tends to be deeply recurved, though in brachiopods with hinged shells there is no anus. The larvae are generally considered to be modified trochophores, so that the lophophorates are embryologically allied with the PROTOSTOMIA. Their characteristics as a whole, therefore, tend to place them between the Protostomia and the Deuterostomia.

It has been suggested by Professor David Nichols in his book *Echinoderms* that the Bryozoa and other lophophorates may be traced to a sipunculid-like ancestor. If so, this would provide a definite link with the Protostomia, for sipunculid early development proceeds by cleavage in the annelid manner to a trochophore which metamorphoses into the adult.

The adult sipunculid shows many of the features to be expected in a bryozoan ancestor. The body is vermiform, unsegmented and provided with an undivided coelom. Anteriorly there is an introvert which terminates in a mouth surrounded by hollow, ciliated, tentacular outgrowths. The alimentary canal is U-formed, with the anus opening near the base of the introvert. The nerve ganglion is dorsal. Excretion is by metanephridia (absent in bryozoa but present in other lophophorates). The body wall incorporates circular muscles

* Most of the metazoan phyla can be placed in one or other of two evolutionary lines. One line embraces flatworms, nemertines, annelids, arthropods and molluscs, and is called the Protostomia; the other line includes the echinoderms, hemichordates and chordates, and is known as the Deuterostomia. The principal distinguishing features concern development. In the Protostomia cleavage is determinate (each blastomere having a fixed fate) and follows a spiral pattern, the mouth arises at the site of the blastopore, the mesoderm forms from a single cell (known as the *4d* blastomere), the coelom is produced by splitting of the mesoderm (schizocoely), and the larva is of the trochophore type. In the Deuterostomia cleavage is indeterminate and radial, the anus arises at the site of the blastopore, the mesoderm and coelom form by enterocoelic pouching, and the larva is not a trochophore. (See, for example, Professor Barrington's *Invertebrate Structure and Function*.)

outside longitudinal muscles, the condition still found in phylactolaemates. The introvert is withdrawn by retractors—presumably derived from the longitudinal muscles of the body wall—of which there may be four, all equal, or the dorsal pair may be shorter; or only the ventral pair may be present. Protrusion of the introvert is effected by contraction of the circular muscles acting through the coelomic fluid.

The number of similarities is impressive, seemingly too numerous to have arisen by coincidence, and near-sipunculid ancestry for lophophorates seems a useful hypothesis.

While discussing closely allied phyla, no mention has been made of the ENTOPROCTA, which were formerly classified with the Bryozoa. The apparent resemblance is in fact entirely superficial.[8] Entoprocts do not have a true coelom, and their tentacles are extensions of the body margin not morphologically comparable with a lophophore. The anus opens inside the tentacular ring, as the name entoproct implies. The more closely the tentacular circlets of the two groups are examined, the more dissimilar they are seen to be. In bryozoa the feeding current enters the top of the bell and leaves between the tentacles; in entoprocts it flows in the reverse direction. In bryozoa particles are swept straight to the mouth; in entoprocts they are trapped in mucus and transported on ciliary tracts. In bryozoa the cilia display laeoplectic metachronal waves (p. 43): in entoprocts they do not.

The entoproct larva is of the trochophore type, but again much modified: it often clearly displays much of the adult organization. In bryozoa there is a cataclysmic metamorphosis followed by the appearance of an entirely new, adult organization. Far from being classes within one phylum, therefore, it is apparent that the Bryozoa and the Entoprocta are not at all closely related.

OUTLINE CLASSIFICATION

The major divisions of the phylum Bryozoa are as follows:

Class I – PHYLACTOLAEMATA

Zooids basically cylindrical, with a horseshoe-shaped lophophore and an epistome. Body wall incorporating muscle fibres; non-calcareous. Mechanism for everting the lophophore dependent on circular muscle in the body wall. Coelom continuous between zooids. New zooids arising by the replication of polypides. Differentiation of the cystids following inception of the polypides. Spermatozoa developing in large masses. Exclusively freshwater; produce special dormant buds called statoblasts. About 12 genera; Mesozoic (?)—Recent.

Class II – STENOLAEMATA

Fossil except for some Cyclostomata (so that the definition of soft parts is based on these alone). Zooids cylindrical. Body wall composed of cell layers, but no muscle fibres; incorporating calcification. Mechanism for everting the lophophore not dependent on muscular deformation of the body wall. Coeloms of adjacent zooids separated by shared walls (septa), though sometimes in communication through open mural pores. New zooids produced in a common colonial growing zone (or zones) by the division of septa; differentiation of the cystids following inception of the polypides. Limited polymorphism of zooids; marine. About 550 genera.

Order 1 – CYCLOSTOMATA

Orifice of zooid circular, devoid of closing apparatus. Lophophore circular; no epistome. Zooids interconnected by open pores. Spermatozoa developing in tetrads. Sexual reproduction involving polyembryony which generally takes place in special reproductive zooids. About 250 genera; Palaeozoic—Recent.

Order 2 – CYSTOPORATA

A recently created order containing certain Palaeozoic genera formerly classified with the Cyclostomata. Zooid skeletons long and tubular, interconnected by pores, and containing transverse partitions (diaphragms); 'cystopores' (not pores but supporting structures between the zooid skeletons) present, also containing transverse partitions. About 50 genera; Palaeozoic.

Order 3 – TREPOSTOMATA

Colonies generally massive, composed of long tubular zooid skeletons having lamellate calcification; without mural pores between zooids; orifices polygonal. Zooid walls thin proximally, thicker distally, sometimes with numerous diaphragms, and interspersed with other calcified structures. About 100 genera; Palaeozoic.

Order 4 – CRYPTOSTOMATA

Colonies mostly with foliaceous or reticulate fronds, or with branching stems. Zooid skeletons tubular, with diaphragms, but shorter than in trepostomes; proximal portions thin-walled and in contact with each other, distal portions funnel-like and separated by extensive calcification. Mural pores absent. About 130 genera; Palaeozoic.

Class III – GYMNOLAEMATA

Zooids cylindrical or squat, with circular lophophore; without an epistome. Body wall composed of tissues, but no muscle fibres; sometimes with calcification. Mechanism for everting the lophophore dependent basically on deformation of the body wall by muscles. Coeloms of adjacent zooids separated by septa or by duplex walls; communicating by pores plugged with tissue. New zooids produced in (often contiguous) branching series behind growing points by the deposition of transverse septa; cystid formation preceding polypide inception. Spermatozoa developing in large masses. Zooids polymorphic. Mainly marine. About 650 genera.

Order 1 – CTENOSTOMATA

Zooids cylindrical to rather flat; walls membranous or gelatinous, not calcified. Orifice terminal or nearly so, often closed by a pleated collar. No ooecia, no avicularia. About 40 genera; Palaeozoic—Recent.

Order 2 – CHEILOSTOMATA

Zooids generally shaped like a flat box, walls calcified. Orifice frontal, closed by a hinged operculum. Specialized zooids, such as avicularia, commonly present. Embryos often developing in special brood chambers known as ooecia or ovicells. About 600 genera; Mesozoic—Recent.

Earlier classifications divided the phylum into two classes: PHYLACTOLAEMATA (*phylasso*, guard+*laimos*, throat; referring to the presence of the epistome) and GYMNOLAEMATA (*gymnos*, naked; referring to the absence of the epistome). The Gymnolaemata were then divided into five orders. The comprehensive study of the Cyclostomata by the Swedish zoologist F. Borg,[119] however, made it clear that the differences between that order and the other extant marine bryozoa (Ctenostomata and Cheilostomata) were considerable, and Borg proposed for the cyclostomes a new class STENOLAEMATA (*stenos*, narrow; taking *laimos* to mean the 'throat' of the zooid rather than of the polypide, referring to the narrow, tubular shape of the walls). Studies on the Trepostomata clearly revealed their affinities with the Cyclostomata, so that they too were classified in the Stenolaemata. Many geologists, but few zoologists, accepted Borg's three class system.

Later studies[85] have emphasized certain similarities between the Ctenostomata and the Cheilostomata, and these seem best recognized by keeping them together in a restricted class Gymnolaemata, but ordinally separate since their differences are still considerable. The fusion of the two orders (as Cheilo-Ctenostomata or Eurystomata) and the retention of an omnibus class Gymnolaemata seems to be the less satisfactory alternative.

Only the Cryptostomata have proved difficult to accommodate in the three class scheme, and they have usually been placed in the Gymnolaemata from the belief that they gave rise to the Cheilostomata; but this cannot possibly be so (Chapter 7). Only recently have some geologists come to recognize their stenolaemate affinities. It has become clear to me, while writing this book, that the characters of the Cryptostomata place them unequivocally in the Stenolaemata. Thus there are three very distinct classes within the Bryozoa. Each must have been in existence by the early Ordovician at least (the Phylactolaemata have left no record) and each has survived as an independently evolving line until the present day.

Although, as the diagnoses show, the Gymnolaemata and Stenolaemata (as exemplified by the living Cyclostomata) have in common several features which are not found in the Phylactolaemata, there are also characteristics shared only between the Gymnolaemata and the Phylactolaemata, or between the Stenolaemata and the Phylactolaemata.

2

GYMNOLAEMATA—
SUCCESS IN THE SEA

The Gymnolaemata are an important invertebrate class, with 3000 or more living species classified into two orders—CTENOSTOMATA and CHEILOSTOMATA—of which the second is by far the larger. This and the succeeding three chapters are concerned mainly with documenting the success of bryozoa in this class: their evolutionary diversity (this chapter); the ways in which they live and interact with their environment (Chapters 3 and 4); and finally (Chapter 5) some of the special features which seem to have contributed most to their success.

CTENOSTOMATA

The order, defined on p. 27, was named Ctenostomata (*kteis*, *ktenos*, comb+*stoma*, mouth) from the belief that the collar, which folds into pleats when it closes the atrium, was a toothed or comb-like structure. There are two suborders: STOLONIFERA and CARNOSA.

Stolonifera

Bowerbankia, described in Chapter 1, is a typical representative of the Stolonifera. It provides us with a simple example of polymorphism, for the *Bowerbankia* colony is constituted from two different types of zooid. In general terms, zooids which perform all the usual body functions can be distinguished from those which are variously adapted for some specific purpose, usually having lost the polypide in the process. The first are called *autozooids*; all the rest are *heterozooids*. The type of heterozooid found in stoloniferans is that which, placed end to end in series and separated by the septa, makes up a stolon and is more specifically known as a *kenozooid* (*kenos*, empty). Each kenozooid is a long, narrow segment lacking internal structure other than the mesenchymatous funiculus, recognizable as a zooid mainly by its wall structure, method of formation and blastogenic ability (p. 58).

In the Stolonifera, the creeping or free branching stolons originate directly from the primary zooid of the colony and bear the autozooids, usually in clusters, at intervals along their length. Two superfamilies, VESICULARIOIDEA and WALKERIOIDEA, are distinguished: in the former (which includes *Bowerbankia* and *Zoobotryon*), the stolons are relatively thick and there is a long budding zone in each kenozooid; in the latter (which includes *Walkeria*), the stolons are very thin, and budding takes place exclusively at the distal end of each kenozooid.

Throughout the Stolonifera the wall between the base of an autozooid and its supporting stolon, like the end walls between stolonal kenozooids, is perforated for the passage of the funiculus which runs through the whole of the colony. It is wrong, therefore, to think of zooids as isolated units.

Carnosa

That it is the colony which possesses individuality can better be seen in the Carnosa. In this suborder, the primary zooid gives rise to autozooids by budding, and each of these is flattish, adherent and not borne on a stolon. *Flustrellidra* and species of *Alcyonidium* are common on the fucoids of sheltered shores around the British Isles, where they form fleshy incrustations on the algal thalli. A piece of *Fucus* may support many colonies. When the growing edges of two colonies of one species meet they do not fuse to form a single colony; neither does one grow over the other as it would over any other flattish obstruction. When two lobes of the same colony meet, however, they unite. Thus we have clear evidence of the individuality of colonies. Carnosan genera with compact colonies of this type are classified in the superfamily HALCYONELLOIDEA.

In the superfamily PALUDICELLOIDEA, the colonies are diffuse, consisting of ramifying chains of zooids. The colonies of some genera are apparently stolonate, but the organizational pattern is quite different from that found in *Bowerbankia*. The 'stolons' are not composed of kenozooids, but are derived from slender outgrowths (*pseudostolons*) from the base of each autozooid. In *Nolella*, for example, the colony comes to resemble a web of pseudostolons with erect, tubular zooid parts appearing to arise from the points where the pseudostolons meet. Two other well-known genera are *Paludicella* and *Victorella* (discussed in Chapter 4).

Study of the variation in zooidal form between species of pseudostolonate and incrusting Carnosa suggests how simply a cylindrical zooid, which is probably primitive in bryozoa, may have evolved into a flat and mainly adnate one. In some species of *Nolella*, and in certain growth forms of *Victorella*, the zooids form tall cylinders. Except that they give rise proximally to pseudostolons, rather than themselves springing from kenozooids, they resemble the autozooid of *Bowerbankia* in form. In other species of *Nolella* there is a distinct lateral dilatation at the foot of the zooid, into which the retracted polypide can partially retreat. In *Paludicella* the cylindrical part of the zooid is very short, but nevertheless contains the longitudinal parietal muscles; the proximal portion is large and contains the transverse parietal muscles and the retracted polypide in its entirety. The connecting regions are short and not stolonal. Some species of *Alcyonidium* have a loose network of rather squat zooids, attenuate towards their points of connexion, but without slender outgrowths. The basal region constitutes most of the zooid, but the orifice is still raised on a papilla. In other species of the same genus the flattened zooids are contiguous throughout, and the orifice lies in the plane of the upper wall. The adherent surface of a flattened zooid is referred to as *basal*, the upper wall as *frontal*. The end nearest to the origin of the colony is *proximal*, exactly as in a cylindrical zooid, and the far end is *distal*. Communication pores, already mentioned as occurring in stolonal septa, are present in all the contiguous walls of adjoining zooids.

One genus, remarkable for being apparently non-colonial, remains to be mentioned. This is *Monobryozoon*,[43,8] the sausage-shaped autozooids of which have been found living solitarily in soft sediments forming the bed of some European seas. It seems that *Monobryozoon* should be considered one of the Paludicelloidea, for adhesive pseudostolons which act as anchors arise from the zooid base in the same manner as in *Nolella* and *Victorella*. The free end of a pseudostolon may swell up and start to form a polypide, and the parent zooid may at any time be associated with one or two such daughter zooids. The colony apparently never gets larger than this, for the parent zooid then dies. Since any pseudostolon is separated from the parent zooid by a septum, even a single functional autozooid without well-developed daughter zooids is still in reality a small colony.

CHEILOSTOMATA

Ctenostomes and cheilostomes (defined on p. 27) are generally assumed to have arisen from a common ancestor. The principal differences between the two groups are, firstly, that whereas the ctenostome body wall is cuticularized and sometimes thickened by non-calcareous material, the wall in cheilostomes is reinforced to a varying degree with calcium carbonate; and secondly, that the orifice is not closed in the same manner. In ctenostomes closure is effected through the combination of a contracted sphincter muscle and the collar—the latter evidently functioning as a one-way valve—or occasionally, as in *Flustrellidra*, by a pair of lips, the lower one of which is movable by muscles. In the Cheilostomata (*cheilos*, lip+*stoma*, mouth), the lip principle has been developed to the extent of forming a single hinged flap, the *operculum*, which fits accurately into an orifice of constant shape. *Bugula*, one of the most familiar genera, is exceptional in that the operculum is secondarily absent.

It is the presence of the operculum which has facilitated another cheilostome peculiarity; the occurrence of specialized heterozooids called *avicularia*—in which the operculum is enlarged to form a 'jaw'—and *vibracula*, in which the operculum has lengthened into a seta. These, together with the characteristic embryo chambers or *ooecia* in which the larvae develop, are discussed fully in Chapter 5.

The shape of the cheilostome zooid is often rather flat and box-like (Fig.3A).

The Cheilostomata are often classified into two 'suborders', ANASCA and ASCOPHORA, representing distinct levels of organization rather than proper taxa. The system has the merit of simplicity but, when examined more closely, can be seen to leave two 'splinter groups' not very satisfactorily attached to either. The scheme adopted in this book, therefore, recognizes four major divisions: ANASCA, CRIBRIMORPHA, GYMNOCYSTIDEA (=Ascophora Imperfecta) and ASCOPHORA (=Ascophora Vera).

Anasca

If the zooid walls have become calcified for better support, how is elasticity retained to preserve the hydrostatic method of lophophore

---

Fig. 3 Zooid structure in Anasca

A Zooid with tentacles expanded
B Transverse section through proximal part of zooid (with the tentacles retracted)
C Transverse section through distal part of zooid (with the tentacles expanded)

For clarity in the diagrams, the walls of adjacent zooids are shown respectively black and hatched. Septa, pores and pore-chambers, and also ooecia, are discussed in Chapter 5.

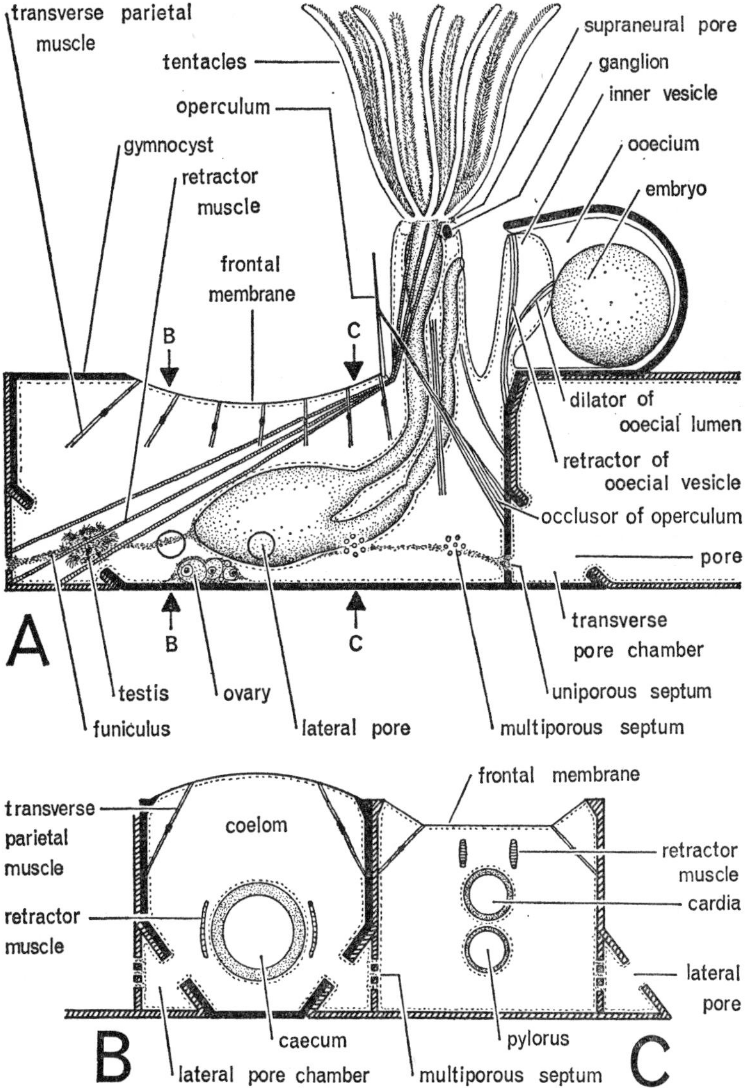
transverse parietal muscle
tentacles
operculum
gymnocyst
retractor muscle
frontal membrane
B
C
supraneural pore
ganglion
inner vesicle
ooecium
embryo
dilator of ooecial lumen
retractor of ooecial vesicle
occlusor of operculum
pore
transverse pore chamber
uniporous septum
multiporous septum
lateral pore
ovary
testis
funiculus
A
B
C
transverse parietal muscle
coelom
retractor muscle
caecum
lateral pore chamber
multiporous septum
frontal membrane
retractor muscle
cardia
lateral pore
pylorus
B
C

Fig. 3

protrusion? The solution, in the first instance, is that only the side walls are calcified and rigid. *Membranipora* shows this clearly, for its lacy appearance results from the contrast of white side walls against the dark algal frond seen through the transparent front of the zooid. This last, now referred to as the *frontal membrane*, has thus the same structure as the wall in *Bowerbankia*.

Modifications to the zooid subsequent to its loss of near-radial symmetry involve mainly the parietal muscles. No longer can each series of transverse muscles be associated solely with the lateral wall; instead each arises from the lateral wall and, spanning the coelom, inserts on the frontal membrane (Fig.3B,C). Contraction of these muscles pulls the frontal membrane inwards, increasing hydrostatic pressure inside the coelom in the manner required.

The operculum, quite evidently on examination, is a hinged fold of the frontal membrane. It is primitively semicircular (Fig.5A,B, operc), and may be stiffened by sclerites around its rim. Its musculature has been developed from the most distal sets of transverse parietals. The operculum is closed by a pair of *occlusor muscles*, one of which arises on each side of the zooid and converges to a tendon which inserts on the operculum (Fig.3A). If these are relaxed, the operculum will be pushed open whenever pressure in the coelom rises; opening may, however, be assisted directly by a pair of small *divaricator muscles* which arise on the lateral walls and insert just behind the hinge line.

Cheilostomes having the basic morphological plan just described are referred to the suborder Anasca (so named for a reason which will become apparent later). A few of the more primitive anascans are placed in superfamilies of their own, but two very large and fundamentally rather similar groups are the MALACOSTEGOIDEA (*malakos*, soft+*stege*, roof; referring to the frontal membrane) and the CELLULARIOIDEA (*cellula*, a little chamber; referring to the cystids). The Malacostegoidea includes *Callopora*, *Electra* and *Membranipora*: they are generally incrusting forms with contiguous zooids, though *Flustra* has erect colonies. The walls, with the exception of the frontal membrane, are calcified. The Cellularioidea includes *Bugula*, *Scrupocellaria* and others with erect, much branched and rather bushy colonies, and having rather lightly calcified walls surrounding the frontal membrane.

Autozooids of many anascans possess spines which project upwards from the calcareous margin around the frontal membrane. *Callopora* (Fig.5A), in which there is a complete ring of spines, appears to be the most generalized example; in many others the spines tend to be restricted to the region of the orifice (Fig.5E).

The origin of the cheilostomes is uncertain. Despite resemblances to the ctenostomes, and the apparent simplicity of the latter, it does not seem possible to find very obvious connecting links among the supposedly most primitive anascans (see, however, Chapter 7). As the zooid of *Membranipora* shows the simplest structure, it was for many years customary to regard it as the most primitive of the Malacostegoidea. Silén,[85] however, produced evidence to show that the *Callopora*-type of zooid is the more fundamental. Also, he believes that the spines which encircle the frontal membrane are not mere projections but a series of kenozooids (see p. 140). In the same connexion, it is noteworthy that in many of the more advanced cheilostomes, with zooids considerably altered from the basic type, the primary zooid or *ancestrula* produced by metamorphosis of the larva exactly resembles the zooid of *Callopora*. The later evolution of the cheilostomes can, moreover, be clearly understood in terms of radiating lines from a basic malacostegoidean zooid like that of *Callopora*. The simplicity of *Membranipora*, it seems, must be explained in terms of loss of spines and certain other characters.

It might be supposed *a priori* that the anascan zooid, with its extensive and delicate frontal membrane, would leave the polypide exposed to predators and to damage inflicted in other ways (see Chapter 4). With such vulnerability in mind, the significance of various evolutionary tendencies seen in the Cheilostomata will readily be appreciated. These trends, indicated diagrammatically in Fig. 4, involve much greater calcification of the zooid without impairing its hydrostatic system.

In the zooid of *Membranipora* the frontal membrane completely roofs the space between the supporting side walls (Fig.4A). In *Electra* and *Callopora*, on the other hand, the frontal membrane is more or less oval in shape, and the part of the zooid proximal to it is covered by a solid front wall, narrow extensions of which also flank the membrane (Fig.5A). This kind of frontal calcification is called a *gymnocyst* (*gymnos*, uncovered+cyst, referring to the cystid or wall of the zooid). Inside the frontal membrane there may also be a narrow, annular interior wall or *cryptocyst* (*kryptos*, hidden), so that the membrane, though still free to move, is reinforced within (Fig.4C^1).

In some genera the cryptocyst may be quite wide, especially proximally. From this condition a line of evolution can be traced which affords increasing protection to the polypide. In *Micropoar* and other genera placed in the COELOSTEGOIDEA (*koilos*, hollow), the cryptocyst has become greatly extended so that it effectively divides the coelom of the zooid into two (Figs.4C^2,5D). A shallow space, called the *hypostegal coelom*, immediately underlies the frontal membrane, while the principal chamber is below the cryptocyst. The proximal transverse parietal muscles appear to have been lost. The more distal bands contribute to a single muscle on each side which, from its origin, converges to a tendon, passes through a small lacuna in the cryptocyst, and inserts on the frontal membrane. The cryptocyst may even extend downwards at the site of the two lacunae, forming calcified cylinders encasing the muscles.

In a second superfamily, the PSEUDOSTEGOIDEA, exemplified by *Cellaria*, there has also been extensive development of a cryptocyst.

Thus in the Coelostegoidea and the Pseudostegoidea the soft parts of the zooid are well protected by a calcified partition which is complete except, mainly, for a space below the operculum. The volume-changing role of the frontal membrane is preserved by the presence of a hypostegal coelom of adequate depth between the membrane and the cryptocyst. The Coelostegoidea appear to be on the decline, being less numerous today than they were in the late Cretaceous and early Tertiary periods.

Fig. 4 Evolutionary trends in the Cheilostomata

A Transverse section through simplest malacostegoidean zooid

B¹ Transverse section through malacostegoidean zooid with marginal spines

B² Malacostegoidean zooid with overarching spines
 a Transverse section through zooid
 b Spines in surface view

B³ Cribrimorph zooid
 a Transverse section through zooid
 b Spines in surface view showing lateral fusions

C¹ Transverse section through malacostegoidean zooid showing slight development of cryptocyst

C² Coelostegoidean zooid
 a Transverse section through zooid
 b Longitudinal section through zooid

D¹ Transverse section through gymnocystidean zooid during development of the frontal wall

D² Transverse section through gymnocystidean zooid after completion of the frontal wall

E¹ Ascophoran zooids during development of the ascus
 a Transverse section of zooid
 b *Camera lucida* drawing of a *Smittina* zooid in frontal view

E² Ascophoran zooids after completion of the ascus
 a Transverse section of zooid
 b *Camera lucida* drawing of *Smittina* zooid in frontal view

a ascus; *crypt* cryptocyst; *f m* frontal membrane; *p m* parietal muscle; *or* orifice.

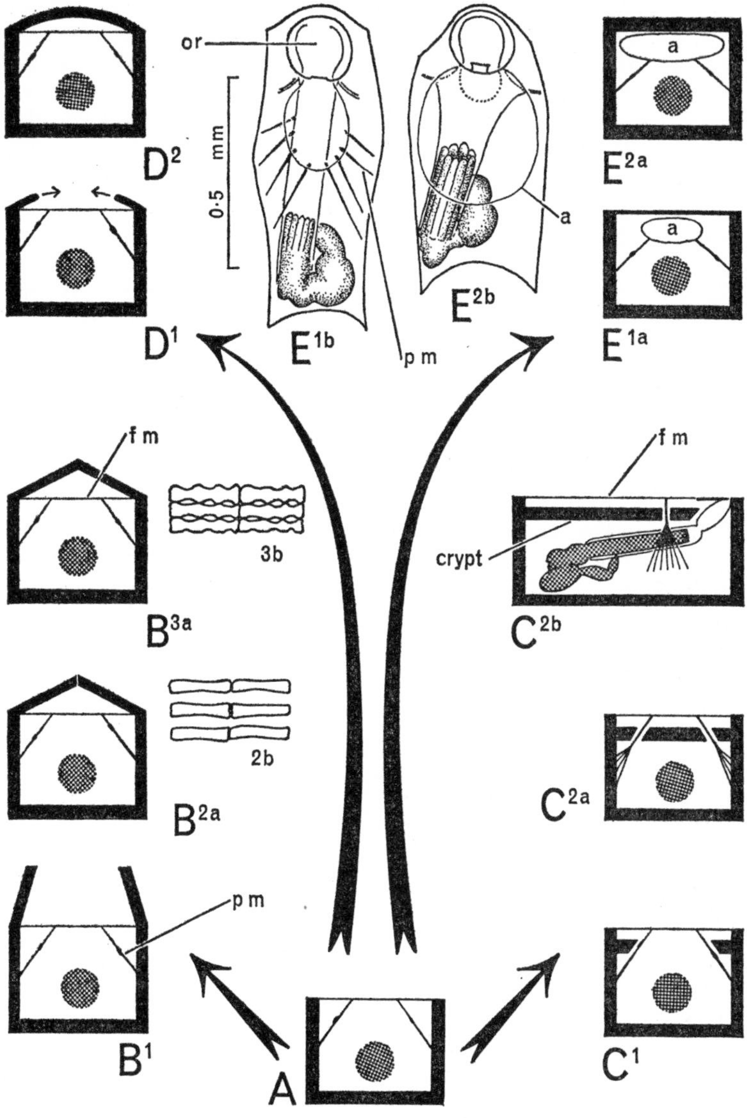

or
0·5 mm
pm
a
fm
3b
crypt
2b
pm
D²
D¹
E¹ᵇ
E²ᵇ
E²ᵃ
E¹ᵃ
B³ᵃ
B²ᵃ
B¹
A
C²ᵇ
C²ᵃ
C¹

Fig. 4

In other anascans the presence of spines around the frontal membrane may afford some slight protection against predators and damage. The actual form and disposition of spines in the genus *Callopora* varies from species to species. In what appears to be the basic type (Fig.4B¹) they extend upwards from the plane of the frontal wall. In another species they are slightly flattened and arch over the frontal membrane like a series of rafters. In a third species they are wide and cross over the frontal membrane parallel to it, fusing in the midline with their opposite number (Figs.4B²,5B); a condition also seen in *Membraniporella*. This particular modification appears to have occurred more than once in cheilostome evolution, most notably when it gave rise to the suborder Cribrimorpha.

Cribrimorpha

In this suborder, which also achieved its zenith during the Cretaceous (p. 141), and is now represented by a handful of species, the spines fuse in the midline, protecting the frontal membrane by a cage of ribs. Adjacent ribs have then merged at intervals along their length, resulting in the formation of a sieve-like shield (Figs.4B³,5C); hence Cribrimorpha (*cribrum*, sieve+*morphe*, form). The pores permit the free passage of water, so that the functioning of the frontal membrane is in no way hampered. The protection provided by a ribbed roofing would seem to represent a clear advance over the malacostegoidean condition yet, while the cribrimorphs have declined, the latter still flourish.

Fig. 5 Cheilostome zooids

A *Callopora* (Anasca Malacostegoidea)
B *Callopora*, another species
C *Cribrilina* (Cribrimorpha)
D *Calpensia* (Anasca Coelostegoidea)
E *Bugula* (Anasca Cellularioidea)
F *Escharoides* (Gymnocystidea)
G *Hippothoa* (Ascophora)
H *Porella* (Ascophora)
I *Schizoporella* (Ascophora)

av avicularium; *az* autozooid; *b b* brown body (see p. 59); *calc f w* calcified frontal wall; *crypt* cryptocyst; *gymn* gymnocyst; *lac* lacuna in the cryptocyst through which passes the tendon of the parietal muscle; *m ps* marginal pseudopore; *m sp* marginal spine; *mod sp* modified spine; *oec* ooecium; *operc* operculum; *or* orifice (drawn black, with the operculum removed, to show the condyles); *prox* proximal part of zooid covered in life (in A) by the ooecium of the next proximal zooid or (in E) by the distal end of that zooid; *ps* pseudopore; ♀ special female zooid; ♂ special male zooid (Certain of the features illustrated will be first fully discussed in Chapter 5).

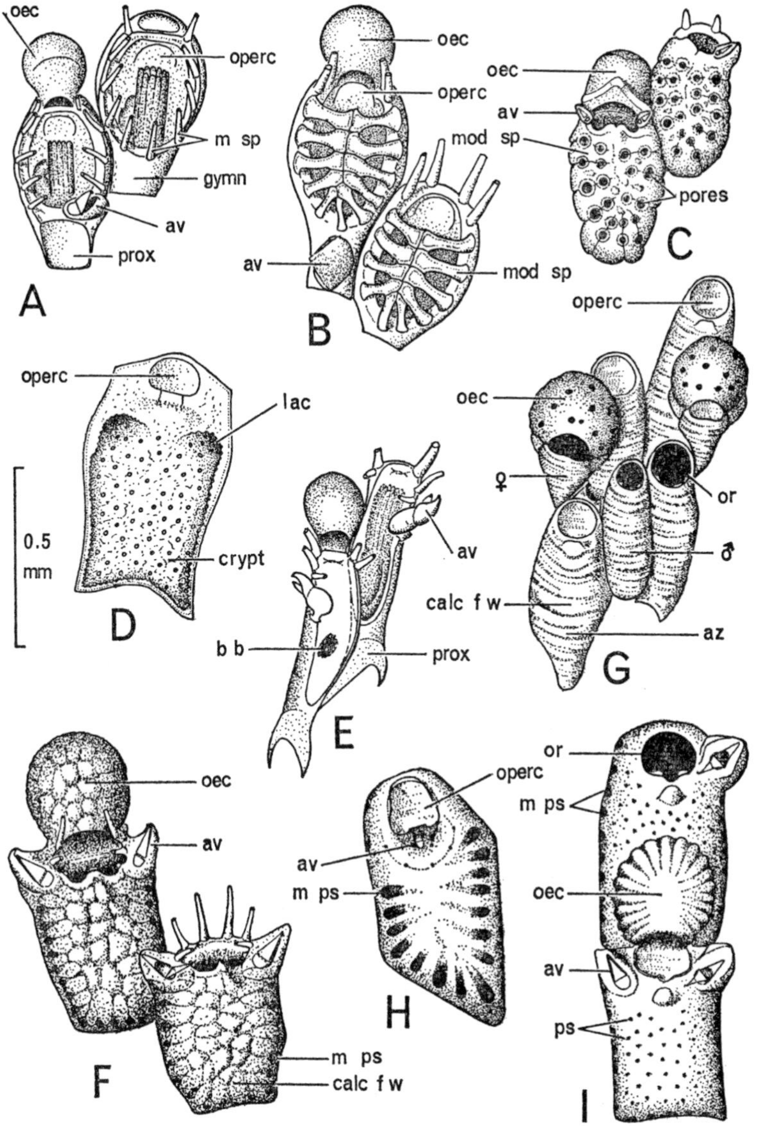
oec
operc
m sp
gymn
av
prox
A
oec
operc
mod sp
av
mod sp
B
oec
av
mod sp
pores
C
operc
lac
0.5
mm
crypt
D
av
b b
prox
E
operc
oec
♀
or
♂
calc f w
az
G
oec
av
m ps
calc f w
F
operc
av
m ps
H
or
m ps
oec
av
ps
I

Fig. 5

Gymnocystidea

The remaining cheilostomes all have the frontal surface of the zooid calcified. This solid wall may form in one of a number of ways. In *Escharoides* and *Umbonula*, for example, developing zooids at the colony margin clearly display the rudiments of anascan structure, in which the gymnocyst or calcified part of the external wall delimits the proximal part of the zooid and also flanks the frontal membrane; but, during ontogeny, an outfolding of the body wall from the margin of the gymnocyst grows upwards and inwards, arching over the frontal membrane (Fig.4D^1). Calcification takes place on the basal side of this fold, at least in *Metrarabdotos*.[5] When the edges meet and fuse the fold has created a vaulted roof above the intact membrane below it (Figs.4D^2,5F). In zooids developing in this way the orifice is D-shaped and there is a wide space above the opercular hinge through which water flows as the frontal membrane functions in the anascan manner. The suborder GYMNOCYSTIDEA was created by Silén[85] for cheilostomes of this kind.

In some genera it seems that a supra-membranal frontal wall, like that just described, has arisen independently from the fusion of ribs comprising the frontal shield of cribrimorphs. All the pores have been eliminated, leaving only an opening by the opercular hinge. Silén[85] classified cheilostomes of this kind as SPINOCYSTIDEA. The present state of our knowledge, however, hardly permits the segregation of Spinocystidea from Gymnocystidea. Thus, without in any way prejudicing future judgments when the subject is better studied, both groups, together with any genera in which the frontal wall arises partly from spines and partly from a fold, are united here as the Gymnocystidea.

Many genera currently classified with the Ascophora (see below) may prove to belong to the Gymnocystidea, but sufficient have already been recognized to vindicate Silén's erection of this suborder and to justify our use of it.

In some of the cellularioid Anasca one much enlarged and shield-like spine or *scutum* may curl over the frontal membrane, as it does in *Scrupocellaria*. It is interesting to find that in a few species of *Notoplites* and *Caberea* the scutum is so enlarged that it forms a tightly-fitting cap over the frontal membrane, paralleling the situation found in the Gymnocystidea. Water again presumably enters alongside the opercular hinge.

Ascophora

In the three suborders so far discussed, hydrostatic control is exercised by means of a flexible frontal membrane. In the Ascophora (*askos*, bag+*phoreus*, bearer) zooids are characterized by the presence of a solid frontal wall, while hydrostatic control depends on an underlying sac, the *ascus*, opening to the exterior at one end. The transverse parietal muscles arise on the lateral walls as before, but now insert on the lower side of the ascus (Fig.4E^1,E^2). Their contraction reduces the coelomic volume in the usual manner, and simultaneously dilates the ascus. The tentacles are pushed out as water enters from outside to fill the ascus and compensate for the decreased coelomic volume.

As originally constituted by Harmer,[50] who gave the first generally accepted account of the ascus and its function, the Ascophora embraced all cheilostomes having a calcified frontal wall, thus including those just described as comprising the Gymnocystidea. In Harmer's last work,[51] eventually published some time after his death, he clearly distinguished the true ascophorans (ASCOPHORA VERA), with an ascus, from those with a covered frontal membrane (ASCOPHORA IMPERFECTA). Meanwhile, Professor Lars Silén in Sweden had independently discovered how the frontal wall forms in the latter group and had introduced the name GYMNOCYSTIDEA. Utilization of this leaves ASCOPHORA freely available for those cheilostomes provided with an ascus, and avoids double-barrelled names. ANASCA, of course, means 'without an ascus'.

The ascus is a delicate structure and not easily discerned below the frontal wall. Nevertheless, in developing zooids of *Smittina* or *Watersipora*, for example, it can be observed through the calcifying wall as it arises from the edge of the orifice (Fig.4E$^{1\,b}$) and enlarges in a proximal direction until fully formed (Fig.4E$^{2\,b}$). The parietal muscles, which radiate from the ascus, may also be conspicuous during its development.

Calcification is apparently laid down in the frontal membrane until an opaque wall covers the zooid except for the orifice and its opercular lid. This method of wall and ascus formation is highly distinctive, and difficult to derive, in terms of evolutionary morphology, from the anascan condition.

The orifice in Ascophora is usually clearly divided into two regions: a larger distal portion or *anter*, which is continuous with the tentacle sheath; and a smaller proximal portion or *poster*, which is continuous with the ascus (see Fig. 15, p. 99). When the poster is narrow, as in *Schizoporella* (Fig.5I), it is termed a *sinus*. Separating the two regions of the orifice is a pair of skeletal protuberances or *condyles*—tiny pointed teeth in *Cryptosula* and *Pentapora* (Fig.15B)—on which the operculum, no longer hinged to the frontal membrane, articulates. When hydrostatic pressure inside the zooid rises through dilatation of the ascus, the operculum pivots on the condyles with its larger part swinging outwards and the smaller part, which closes the poster, dipping inwards. The movement which allows the tentacles to emerge, simultaneously facilitates entry of a compensating flow of water into the ascus.

Typical genera in the Ascophora with bipartite orifices (Fig.15A,D) are *Hippothoa*, *Schizoporella*, *Smittina* and *Watersipora*. In *Microporella*, the external opening of the ascus has become separated from the orifice, and is seen as a median pore (Fig.15C); and the orifice is semicircular, not bipartite. *Haplopoma* is interesting in this connexion, for the ancestrula has a *Schizoporella*-like type of orifice, whereas in the subsequent autozooids the ascus opens through a separate pore (*ascopore*) in the manner of *Microporella*.

The frontal wall in ascophorans, as it first forms, is apparently still inadequate to protect the polypide, for a great majority of species strengthen it by further calcification added to its outer surface. This *secondary thickening* may be continued until the wall assumes relatively massive proportions, accounting for the coral-like texture of *Pentapora* for example. The processes involved are discussed in Chapter 5.

3

PHYSIOLOGY—
FUNCTION IN ZOOID AND COLONY

NUTRITION

The lophophore and food capture

Bryozoans catch planktonic food with the aid of a circlet of ciliated tentacles. The expanded tentacles fan outwards from their origin on the lophophore to form a funnel with the mouth at its vertex (Fig.6A). The distal end of each tentacle curves outwards a little; but the proximal region may be held fairly straight, as in cyclostomes, or be convex, so that the funnel is bell shaped, as in many cheilostomes and ctenostomes.

Each tentacle is roughly rectangular or triangular in section, in the second case with the apex centripetally directed (Fig.6C). The outer surface of each is unciliated, but each side bears tracts of long *lateral cilia*. The cilia beat obliquely downwards and outwards in relation to the long axis of the tentacle and exhibit metachronal waves which pass up the left side of the tentacle (viewed from inside the cone) and down the right side: the effective beat is thus to the *left* of the direction of travel of the waves, which are therefore, in the terminology devised by Professor Knight-Jones, *laeoplectic*. The combined action of these cilia is to generate a water current which enters the top of the bell and passes outwards between the tentacles.[26] Some of the particles in suspension must get carried away by the outflowing water, but an adequate proportion will be directed straight at the mouth. Bryozoa, and other lophophorates similarly, may thus be described as *impingement feeders*.[9]

The inner surface of each tentacle is also ciliated, though these *frontal cilia* are shorter and more sparsely developed than the lateral ones. Towards the base of each tentacle the frontal cilia may be more strongly developed, and their function must be to augment the main water current and to direct it towards the mouth. This effect is enhanced by the beating of cilia inside the pharynx, so that smaller particles are drawn in through the mouth. Larger particles are ingested by the sudden dilatation of the lower portion of the pharynx, which thus acts like a suction pump.

Unwanted particles may be prevented from reaching the mouth by individual or concerted movements of the tentacles; closure of the mouth, which results in particles being carried away by the outflowing current; or ciliary reversal which ejects particles from within the pharynx. In *Flustrellidra* and many other ctenostomes, and in at least *Bugula neritina* and *Watersipora*[9] among the cheilostomes, there is a definite ciliary rejection tract leading from a groove inside the pharynx across the lophophore between the most ventral (abneural) pair of tentacles (Fig.6B,rt).

Suspension feeders in the sea are believed to feed mainly on phytoplankton, although bacteria and detritus, for example, may also be utilized, especially in the deep sea. Little is known about the kind and numbers of organisms captured by bryozoans, although experiments and culture work have indicated at least some of the species which are suitable as food. The chrysophycean flagellate *Monochrysis* and the diatom *Phaeodactylum* are good foods for *Zoobotryon*, and some growth occurs with the coccolithophore *Cricosphaera*. Bullivant[35] found no growth with the chlorophycean flagellate *Dunaliella* or with the dinoflagellate *Amphidinium*. *Bugula*, *Conopeum* and other bryozoans have, however, been successfully cultured by Schneider[80] and Jebram[1] on the dinoflagellate *Oxyrrhis* reared on *Dunaliella*.

An aspect of feeding which has so far received no attention is the influence of the dimensions of the tentacular funnel on the type and size of prey captured. On British shores *Flustrellidra* and *Alcyonidium* occur together on the alga *Fucus serratus*, but the tentacular bell of the former (about 1·4 mm diameter and 1·1 mm high) is a great deal larger than that of *Alcyonidium* (about 0·8 mm diameter and 0·6 mm high). It may be that the two species are competitors for space, but that they are taking different food organisms.

The standing crop of phytoplankton in the sea perpetually fluctuates within wide limits, and estimates of numbers of cells present at any locality may vary in time by an order of magnitude or more. The values are usually highest in coastal waters. Typical ranges[35] for spring and summer (in cells/ml) are: diatoms 50–500, dinoflagellates 20–50, coccolithophores 20–100 and flagellates <100–>5000.

The performance of a suspension feeder can be expressed in terms of two parameters: the quantity of food ingested and the volume of water filtered clear. These have been investigated by Bullivant[35] for the warm-water ctenostome *Zoobotryon*. When fed on *Monochrysis* at 23–25°C the maximum ingestion rate was about 3000 cells/zooid/hr achieved in food concentrations of 7900 cells/ml or more. This value, the lowest concentration of particles at which the animal achieves a maximum ingestion rate, is the *satiation concentration*. For *Phaeodactylum* this concentration was 6000 cells/ml, and for the much larger *Cricosphaera* 1700 cells/ml. Such concentrations, it will be noted, are substantially above those occurring naturally in the sea except, perhaps, at the time of algal blooms.

Below the satiation concentration, *Zoobotryon* filtered clear a suspension of *Monochrysis* at a rate of 0·33–0·40 (mean 0·368) ml/zooid/hr. This is equivalent to 33·7 ml/mg dry wt/hr, and compares

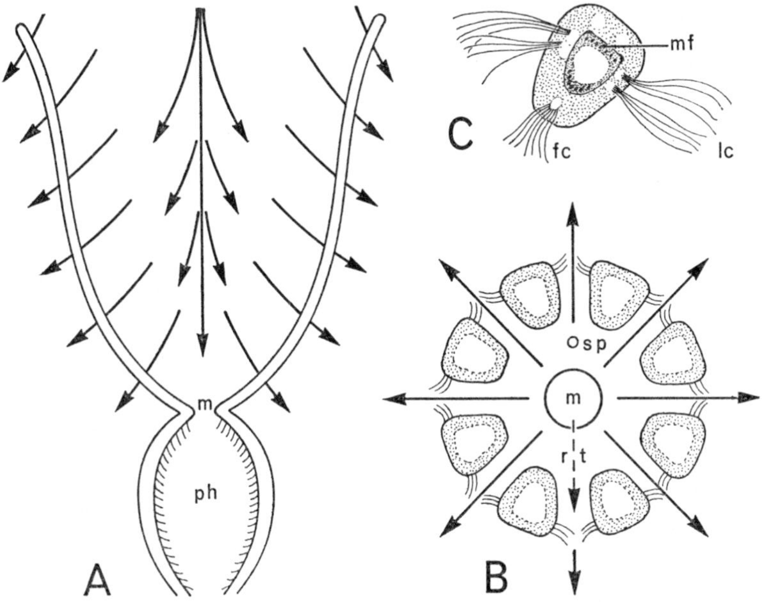

Fig. 6 Feeding in bryozoans

A Longitudinal section through tentacles, mouth (*m*) and pharynx (*ph*)
B Transverse section through the tentacles just above the mouth (*m*), showing also the position of the supraneural coelomopore (*s p*) and of the rejection tract (*r t*)
C Transverse section through a tentacle: *f c* frontal cilia; *l c* lateral cilia; *m f* muscle fibre.
Solid arrows indicate direction of water flow.
(After Atkins[26] and Borg[119])

with values of 17 in the sponge *Sycon*, 7·6 in the polychaete *Pomatoceros*, 6·6 in the oyster *Ostrea* and 56 in its veliger, and about 1 in the ascidian *Phallusia*. In general clearance rates tend to be proportionally higher in small animals. Although the colony of *Zoobotryon* may be quite large (tufts up to 0·5 m long), the zooids are small and evidently, in this particular process, function as independent organisms.

Alimentary canal and digestion

The alimentary canal[87] in each zooid forms, in its entirety, from the tissues of the body wall: there is no midgut of entodermal origin. The anterior region of the gut has no digestive function. In all bryozoa it is possible to distinguish a ciliated region of the pharynx, just inside the mouth, and a non-ciliated region (sometimes called the oesophagus) which generally merges with the former without narrowing; it terminates with a valve or constriction between itself and the stomach.

The stomach, which is differentiated into three regions, has walls which are secretory and absorptive. The first section or *cardia* is shortly tubular in species with cylindrical zooids, but is longer in those with short broad zooids in order to allow the retracted tentacles to lie alongside, rather than above, the alimentary canal. The middle sac-like part of the stomach is the *caecum*, and the ascending lobe is the *pylorus*. In Gymnolaemata and Stenolaemata, but not in Phylactolaemata, the pylorus is ciliated. A pyloric constriction separates the stomach from the posterior part of the gut. The latter may be externally divisible into regions termed intestine and rectum, but there is no evidence for food absorption in the former. The anus opens outside the lophophore near the origin of the most dorsal pair of tentacles.

In certain Gymnolaemata, such as *Bowerbankia* (Fig.2) and *Zoobotryon*, that part of the cardia adjacent to the caecum is transformed into a *gizzard*. This is a spherical muscular organ whose lining epithelium bears hard denticles, usually one to each cell.[29] The gizzard presumably crushes the skeletal cases of diatoms, but few studies have been made on the diet of bryozoa with and without a gizzard. Bullivant[9] found that *Zoobotryon* flourished on a diet of the diatom *Phaeodactylum*, whereas *Bugula*—lacking a gizzard—did not.

Food, having passed through the pharynx, is carried down into the caecum by a peristaltic wave or by total contraction of the cardia. Here mixing takes place by muscular action. Both cardia and caecum are lined by glandular epithelium, and extracellular digestion takes place at pH 6·5–7·0. Particles are formed into a rotating cord by the action of the epithelial cilia in the pylorus. This cord projects from the pylorus into the caecum and, in *Membranipora*, rotates 70–150 times a minute. Edible particles are engulfed by cells of the epithelium in all three parts of the stomach, and an accumulation of brownish particles is characteristically seen in these cells in the cardia and caecum. Starch grains fed to *Zoobotryon* were hydrolyzed in the stomach lumen and absorbed as soluble products. These later re-formed as glycogen, which was observed in the digestive epithelium and in the main funiculus of the zooid, passing in time through the basal pore into the stolonal funiculus. Fat and protein are probably also distributed in this manner.

Muscular movements of the pylorus force the compacted remains of the rotating cord into the rectum, through which faeces are transferred to the anus. In *Zoobotryon* fed on *Monochrysis*, passage of food through the gut at 22°C took about 1 hr.[35]

RESPIRATION

Bryozoan zooids are fairly small and it has generally been assumed that oxygen reaches the internal viscera by diffusion through the body wall. In species having some or all of the walls uncalcified this generalization appears true. Mangum and Schopf[62] found that colonies of the anascan *Bugula* with tentacles expanded consumed 3·7 μl oxygen/mg non-skeletal dry wt/hr or $4{\cdot}25\times10^{-3}$ μl oxygen /zooid/hr at 20°C. At this temperature sea water contains about 5·2 ml oxygen/l. In an analysis of their results, which has subsequently been criticized (see below), the authors concluded that this requirement could not be met by diffusion through the frontal membrane alone. They thought that the oxygen-rich coelomic fluid from the lophophore and tentacle sheath would be circulated to the lower part of the metacoel by the recurring process of rapid withdrawal of the tentacles followed by their immediate re-emergence.

Of the $4{\cdot}25\times10^{-3}$ μl of oxygen required hourly by the whole zooid, it was estimated that $3{\cdot}7\times10^{-3}$ μl/hr would be utilized by the cystid and viscera within it. This may well be an overestimate, because the oxygen consumption of ciliated tentacles is likely to be above average, and that of the viscera below average; but the figure can be accepted as a maximum. The formula applied to these results by Teal[95] is derived from that describing diffusion into a cylinder. When modified to fit the present situation, in which oxygen can enter only through the frontal membrane, this is:

$$C=\frac{Ad}{D},$$

where C is the concentration of oxygen in ml/ml sea water at the surface of the zooid, A is the rate of oxygen consumption by the tissues in terms of unit volume and time (ml/ml sec), d is the depth of the zooid from the frontal membrane in cm; and D is the diffusion coefficient (averaged through an appropriate combination of tissue and coelomic fluid) in cm^2/sec. Solving the equation for A, Teal calculated that oxygen could reach the viscera at a rate of $20{\cdot}0 \times 10^{-3}$ μl/hr, or roughly five times the amount required.

Massaro and Fat[66] used Mangum and Schopf's data to develop a different model. They assumed that the *Bugula* zooid could be represented as a hemi-cylinder whose curved surface only was permeable to gases. This, in fact, is more nearly the converse of the truth (cf. Fig.3B,C), and the area through which oxygen can diffuse into the zooid is $\pi/2$ times less than in their model. Nevertheless, this probably has little effect on the general validity of their conclusions. They derive an equation for a diffusion system containing a chemical reaction, assuming that oxygen removal occurs in the tissue layers only. Their approach, which has the merit of separating diffusion in the respiring tissues from that in the non-absorptive coelomic fluid, shows that the oxygen concentration inside the zooid never falls below about 4·8 ml/l or 93% that of saturated sea water at 20°C.

It is quite clear, therefore, that the oxygen requirements of a *Bugula* zooid are easily met by diffusion through the frontal membrane, although this does not disprove Mangum and Schopf's suggestion that benefit might accrue to the zooid from the circulation of coelomic fluid in the manner they propose. What happens, though, when all the zooid walls are calcified? Oxygen requirement will be highest when the lophophore is extended for feeding, but then both tentacles and tentacle sheath will be in direct contact with the sea water. In the Cyclostomata the calcification in the zooid wall is perforated regularly by *pseudopores* filled with living tissue. It has generally been assumed that the function of these is to permit the passage of dissolved gases through an otherwise impermeable barrier, and Borg[119] drew attention to the absence of these pores in the vicinity of the readily permeable membrane which closes the terminal orifice. The brood chambers bear about twice as many pseudopores as the autozooids per unit area of surface: clearly, this must be to satisfy the high oxygen requirement of the developing embryos.

In Ascophora the respiratory adaptations are rather different. The calcification in the frontal wall may contain perforations, but they are often few and confined to the margin of the wall (Fig.5H): the principal function of such pores is probably the facilitation of secondary calcification (p. 97). The ascus, however, being a thin-walled sac opening to the exterior, is perfectly adapted to function as a respiratory organ. Each time the lophophore is retracted and protruded, the ascus will be emptied and refilled with fresh, fully oxygenated sea water.[77]

REPRODUCTION

Most gymnolaemate zooids are hermaphrodite. Occasionally, however, individual zooids within the colony are differentiated as female, as in the cyclostomatous genus *Crisia* (Fig.17A,gz, p. 107), or male, as in *Hippopodinella*.[46] In *Hippothoa* (Fig.5G) and *Thalamoporella* male, female, and sterile autozooids are all present. Germ cells differentiate from the peritoneum of fertile zooids and give rise to ductless gonads on the body wall, polypide or funiculus (Figs.2F,3A). The ovary, a cluster of oocytes contained by a peritoneum, is generally more distally situated than the testis in a hermaphrodite zooid, and may develop later.

Gametogenesis

In gymnolaemates which brood their embryos (the majority), one oocyte at a time enlarges and finally bursts out of the containing peritoneum. After ovulation another oocyte starts to develop, although its growth is held back so long as the first ovum remains in the coelom. The fully developed egg commonly measures about 0·2 mm in diameter. A few anascans, especially *Conopeum*, *Electra* and *Membranipora*, and some ctenostomes, on the other hand, produce many tiny eggs; and these pass directly into the sea.

Spermatogonia derived from the peritoneum proliferate to produce morulae of spermatocytes clustered around a shared cytoplasmic mass or *cytophore*. Meiotic division of the spermatocytes gives rise to spermatids, and finally to morphologically mature spermatozoa which break away from the cytophore. Each spermatozoon is filiform in shape: the head is slender and pointed, the middle region is greatly elongated, and the tail rather short, thread-like or lanceolate with an axial filament (Fig.7A).

One notable feature in spermatid development is that the mitochondria accumulate in a quartet of spheres disposed around the proximal end of the axial filament. Later, as the plasm from the divided spermatocyte migrates down the axial filament, creating the middle region of the sperm, the spheres disintegrate and the mitochondria are distributed with the plasm. Franzén[13] regards the presence of four mitochondrial spheres in mature sperm as a primitive feature, i.e. one associated with fertilization achieved by the discharge of gametes into the surrounding water. Other primitive features are a short middle region and a very long tail: in these respects bryozoan sperm are clearly much modified, so it is interesting that there should be even a transitory appearance of the quartet of mitochondrial spheres.

Fertilization

The mechanism of fertilization in bryozoa has long puzzled zoologists who, having failed to witness the event, have been left with the unsatisfactory hypothesis that self-fertilization within the zooid must generally occur. At last, however, in 1966 Professor Lars Silén[91] discovered how cross-fertilization takes place.

Mature spermatozoa occur in great numbers in the metacoel. In *Electra* their route out of the body is through the dorsal opening leading into the mesocoel or ring coelom of the lophophore, and then up the lumina of the two most dorsal (adneural) tentacles to escape through a minute terminal pore in each. As many as 2000 spermatozoa were estimated to have left one *Electra* zooid in an hour by this route. Sperm release via the tentacles has been reported subsequently[34] for the ascophoran *Schizoporella* and the ctenostome *Zoobotryon.*

Silén observed that the liberated spermatozoa in *Electra* were at once caught up in the water current flowing out of the tentacular bell, and thus carried away from the parent zooid until drawn towards the crown of another zooid where they clung to the tentacles.

In species which shed their eggs direct to the sea through the supraneural coelomopore, or its tubiform modification the *intertentacular organ*, fertilization occurs during the discharge of ova. The timing of fertilization in brooding species is unknown, and the aggregation of sperm has not been observed, but the egg still leaves the zooid through the coelomopore.

Brooding

The brooding of embryos takes place very rarely in the coelom itself (*Nolella*, *Victorella*); more frequently in ctenostomes (*Alcyonidium*, *Bowerbankia*, *Flustrellidra* and *Walkeria*) the large, yolky egg is extruded through the supraneural pore to develop within the introvert (Fig.2G), while the polypide degenerates. In a few other ctenostomes, in the primitive anascan *Labiostomella*, and in the ascophorans *Cryptosula* and *Watersipora*, development proceeds in an invagination of the atrial wall or internal ovisac. Other primitive anascans (*Aetea*, *Eucratea*)[42] have external ovisacs. The more usual ovicell or *ooecium* of cheilostomes (Figs.3A,5A,B,C,E,F,G,I) may perhaps be an extension of the ovisac arrangement in which the outermost layer (*ectooecium*) has become calcified (see Chapter 5).

Silén[88] has described how an egg is transferred from the coelom into the lumen of the ooecium. Movements within the zooid bring the ovum into the upper part of the tentacle sheath of the protruded polypide, immediately below the supraneural pore. The polypide is then partially withdrawn into the cystid so that the pore lines up with the opening of the ooecium. Presumably the *inner vesicle* which normally seals off the ooecial lumen is pulled downwards by its retractor muscles (Fig.3A). The large yolky egg then runs out rapidly as a fine thread through the supraneural pore into the ooecium, where it re-forms as a spheroid. Transfer takes about 15 seconds.

In *Tendra* a remarkable situation occurs. The eggs are small but, instead of being shed into the sea, they are passed into a brood space between the frontal membrane and a series of ribs formed from overarching spines (as was described on p. 38 for *Membraniporella*), where they develop.

Embryology

The vitelline membrane can be observed around the egg immediately after its arrival in the ooecium, and the first cleavage division occurs soon afterwards. The embryo is very yolky and often coloured yellow, orange or red, probably by carotenoids[71] derived from the zooid's food: sometimes it is an intense white. Early development, however, is not greatly affected by the presence of yolk, and proceeds alike in brooded and non-brooded embryos.

Cleavage is total and generally equal; its pattern is radial, forming tiers of cells in line with each other, until a slightly flattened blastula results. Gastrulation begins when four cells at the vegetal pole divide off their inner halves into the blastocoel, so initiating the production of entoderm. The four cells proliferate and form a mass of cells which almost completely fills the blastocoel; some of the cells become mesenchymatous. In brooded embryos, a more or less equatorial ring of enlarged cells acquire cilia and constitute the *corona* or locomotory organ of the larva. These cells elongate until they cover most of the embryonic surface (Fig.7D).

At the animal or aboral pole, the ectoderm differentiates into a sensory apical organ delimited by a groove, whilst two invaginations can be discerned on the oral side of the corona. The anterior of these is the glandular *pyriform organ*, which is closely associated with a cleft, the *ciliated groove*, at the end of which is a tuft of long *plume cilia*. The posterior invagination is large and deep and constitutes the *adhesive sac*, which serves to fix the larva at the onset of metamorphosis.

Larval biology

In *Callopora* the larva leaves the ooecium about a fortnight after the entrance of the ovum.[88] The inner vesicle, which closes the ooecium, is pulled down, and the larva struggles free by its own exertions. The escape of the larva appears to be a response to illumination, and its behaviour thereafter is greatly influenced by light. In all but a few shore species,[73] larvae at first display positive phototaxis but develop a negative response before settling. This accords with the situation found in the larvae of most marine invertebrates. The changeover appears to be related to the passage of time rather than to light energy received, and is more quickly achieved at higher temperatures.[23]

In addition to phototaxis, it seems probable that bryozoan larvae display photokinetic responses. Settlement in shaded places, especially on the lower surface of experimental panels in the sea, has been widely reported in the literature. Although this has quite often been attributed to the deterrent effect of accumulations of silt on the upper surface, rather than to positive attraction to the lower surface, the same settlement pattern occurs in the absence of silt and is reversed when the panels are lit from below.[73] Aggregation in a shaded area may, therefore, perhaps best be explained in terms of low photokinesis, but further work is required on the role of taxes and kineses during the behaviour leading to settlement.

Fig. 7 Reproduction and growth

A Spermatozoon of *Flustra* (after Franzén[18])
B Cyphonautes larva of *Membranipora;* arrows indicate the feeding current
C 'Giant buds' and zooids from the growing edge of a *Membranipora* colony (after Lutaud[60])
D Larva of *Bugula* (after Calvet[36] and Lynch[61])
E 'Umbrella stage' during the metamorphosis of a *Bugula* larva (after Lynch[61])
F Section through a *Bugula* ancestrula 3 hours after fixation of the larva (after Calvet[36])
G Fully formed *Bugula* ancestrula and first bud

add adductor muscle; *ad sac* adhesive sac; *a s o* apical sense organ; *ax fil* axial filament; *ect* ectoderm; *ev ad sac* everted adhesive sac; *ex* exhalent mantle chamber; *in* inhalent mantle chamber; *mid* middle region of spermatozoon; *perit* peritoneum; *ph* pharyngeal funnel; *pl* plume cilia; *p o* pyriform organ; *pol rud* polypide rudiment.

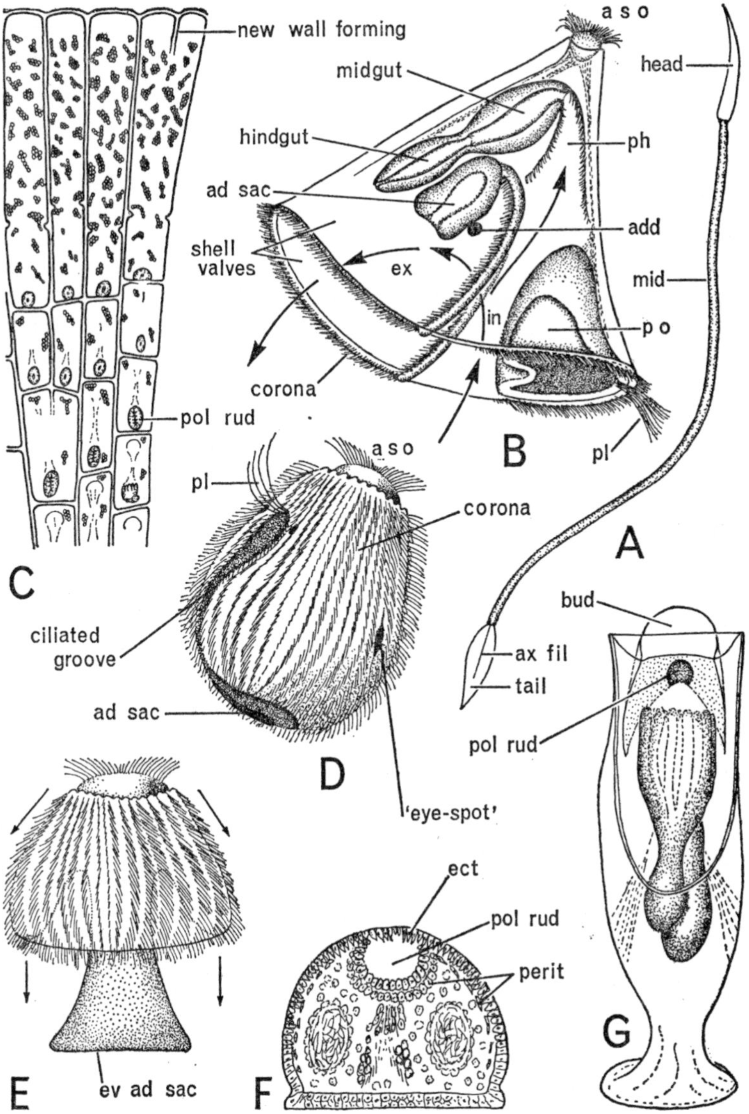
new wall forming
midgut
hindgut
ad sac
shell
valves
ex
in
corona
pol rud
aso
head
ph
add
mid
p o
B
pl
aso
pl
corona
A
C
ciliated
groove
ad sac
D
'eye-spot'
bud
ax fil
tail
pol rud
ect
pol rud
perit
G
E
ev ad sac
F

Fig. 7

Metamorphosis

Since there is no alimentary canal the larva cannot feed, and settlement occurs after a few hours of free existence. The larva selects a suitable spot for attachment (discussed further in Chapter 4) with the aid of the plume cilia. The adhesive sac is then suddenly everted by muscular contractions; it spreads over the substratum as a disc, adhering to it by a secretion coming in part from the pyriform organ. The apical part of the larva moves downwards during the 'umbrella stage' (Fig.7E), enclosing an annular cavity. All the larval organs are absorbed and undergo histolysis, and the annular space becomes merged in the general cavity of the interior of the larva; the ectoderm remains and secretes a cuticle. This stage is reached within about one hour from the onset of metamorphosis.

Cyphonautes

In *Electra*, *Membranipora* and other bryozoans which shed their eggs into the sea,[145,150] the embryo develops into a triangular larva known as a *cyphonautes* (thought, when first discovered, to be a rotifer and named *Cyphonautes*: the name remains as a descriptive label). The larva (Fig.7B) is compressed between the two parts of a bivalved shell, which are held together by an adductor muscle. The apex or aboral pole—which is generally directed forwards as the larva swims—is surmounted by a sensory organ from which nerve fibres radiate; while the base, where the shell valves gape, is encircled by the corona. The direction of ciliary beat is from the apex downwards, while the metachronal waves travel in a clockwise direction when viewed from above and are, therefore, laeoplectic.

The midgut forms from entoderm and merges with the stomodaeal and proctodaeal invaginations so that there is a functional, tripartite alimentary canal. Within the corona is an extensive mantle cavity, bounded in the fully developed larva by the pyriform organ anteriorly and the adhesive sac posteriorly. Two ciliated lobes or ridges, between which the feeding current flows, divide the mantle cavity into inhalent and exhalent chambers. Apically the inhalent chamber merges with a ciliated pharyngeal funnel, which proceeds via a constriction to the midgut; a short hindgut leads to the exhalent chamber. Lateral cilia on the two mantle ridges create the water current;[27] small phytoplankton organisms separated from the water travel up each ridge on its tract of frontal cilia until the pharynx is reached. It appears that food in the midgut is formed into a rotating cord by cilia bordering the pyloric constriction.

The larva grows considerably during its life, until its length exceeds 0·8 mm in *Membranipora*; it then settles and becomes fixed.

GROWTH

The ancestrula

When the larva settles it metamorphoses into a primary zooid or *ancestrula*. This at first comprises a dedifferentiated mass within the body wall (i.e. it is a cystid only), but in most gymnolaemates a functional polypide develops within 2–4 days. Its rudiment appears near the centre of the outer surface of the cystid. Cells of the ectoderm proliferate and form a vesicle which pushes away from the wall into the interior of the zooid (Fig.7F). Other cells, said to be mesectodermal, cover the vesicle to form the peritoneum. The distal part of the bud gives rise to the tentacle sheath, whilst the proximal part forms the rest of the polypide. A small invagination produces the ganglion, and the muscles develop from mesenchyme cells. The orifice arises from a thickening in the cystidial ectoderm which becomes united with the tentacle sheath.

Thus in most of the Gymnolaemata the ancestrula is a complete zooid (Fig.7G), though in some of the Stolonifera a polypide never develops. The ancestrula of *Membranipora* is unusual in that it consists of twin zooids.[145] The external form of the ancestrula may differ from that of the daughter zooids to which it gives rise, especially in the Ascophora. In this group it frequently resembles an anascan zooid of the type found in *Callopora* and is known as a *tata*, or it at least displays a less highly evolved structure than the daughter zooids (as in *Haplopoma*, mentioned on p. 42, or by lacking secondary thickening).

Budding

The ancestrula initiates colony formation by the asexual production of daughter zooids. Each of these is created by the formation of a partition within the structure of the primary zooid, or across a protuberance arising from it, so that one or a series of chambers results. This process of proliferation continues according to a pattern which characterizes the species.

Astogeny

A bryozoan colony displays at least two astogenetic phases (astogeny, from *asty*, town, is the development of a colony) distinguished by budding pattern and zooid morphology. The first phase, described by Boardman[5] as the *stage of astogenetic change*, represents the founding of the colony. The first budded zooids are characterized by morphological variation which usually follows a gradient of increasing size and complexity. Such zooids, to avoid ambiguous use of the word young, may be described as *neanic* (*neanikos*, youthful).

As colonial growth proceeds away from the ancestrula, morphologically comparable zooids appear in one or more endlessly repeatable patterns of budding, so constituting the *stage of astogenetic repetition*. These zooids may be termed *ephebic* (*ephebos*, one who is adult) to avoid confusion with ontogenetic age or maturity. Ontogenetic changes can be inferred from the morphology of zooids at increasing distance from the growing edge of the colony; astogenetic changes can be inferred from the morphology of zooids at increasing distance from the ancestrula.

In a cheilostome ancestrula there may be a single distal bud (Fig.7G), which, with continued multiplication in the same direction, produces narrow rami with zooids arranged in one to several series (commonly two, as in many cellularioids). In such forms branching may be achieved simply by division of a multiserial ramus; but often there is a patterned arrangement of zooids repeated in mirror image at consecutive bifurcations. In incrusting species there may typically be one, two or three distal buds cut off from the ancestrula (as in Fig.12C, p. 89), or five in a lateral-distal-lateral arc, or a complete ring of six; the very young carnosan colony consists of the primary zooid surrounded by a ring of chambers. These daughter cystids enlarge and form their own polypides in the same way as was described for the ancestrula, except that the polypide rudiment usually arises at the proximal end of the frontal body wall (Fig.7C,G). The first generation of daughter zooids produces further buds, and so on until, in many cases, a more or less circular colony is formed, wholly or partially surrounded by an active growing edge. Zooids at the periphery produce either a single distal bud or a pair of buds to allow for the ever-increasing circumference.

Erect colonies may arise directly from an upright ancestrula, as in *Bugula* (Fig.7G) and other cellularioids; or the erect zooids may grow upwards from an initial crust, as in *Flustra*.

The Stolonifera depart from the described pattern in that the ancestrula gives rise only to stolons. Autozooids arise at intervals along the stolons.

The growing edge

Detailed studies of the growing edge have been made in *Membranipora* (Fig.7C) and *Bugula*. In the incrusting colonies of the former, studied by Dr Geneviève Lutaud,[60] juxtaposition of zooids at the edge suppresses lateral budding,[7] and a growing area of limited extent is formed, which progresses rapidly forwards. The colony wall at the margin of this zone comprises cuticle, epithelium and double peritoneum. The same layers occur in both frontal and basal surfaces, but are thinner in the former.

In *Membranipora* the epithelial cells are columnar at the actual edge of the growth zone; in *Bugula*, where growth occurs at the apices of the rami, the epithelium is there replaced by a more or less isolated group of spherical cells.[80] In both genera the marginal or apical cells extend the cuticle by intussusception: that is by the interpolation of new material among the cuticular elements already present. The apical mass of secretory cells displays regular side to side oscillations, which appear to be necessitated by this method of cuticle formation. These cells do not divide, but are pushed forward by a growth zone behind them. New secretory cells may, however, be recruited from the epithelium in the growth zone. In effect, the cuticle is 'stretched' to allow the colony it contains to spread. Mitotic counts and the use of marks of vital stain have shown that in *Membranipora* cell division and growth of the colony occur together in the abapical part of the marginal zone, behind the line of cuticular expansion.

As morphogenesis proceeds, some cells of the inner peritoneal layer of the basal wall are utilized in the formation of the two longitudinal funiculi which characterize this species, whilst corresponding cells from the frontal wall contribute to the developing tentacle sheath. Existing longitudinal walls are continued throughout the entire marginal zone, but new walls, at the point of division of a zooid row, are initiated among the apical palisade cells (Fig.7c). Cystid formation continues with the creation of a transverse wall behind the growth zone. This arises as an annular invagination of the epithelium, initiated at the base of each lateral wall. Multiporous communication plates are produced where the two funiculi transect the developing wall. Meanwhile, in a median position just distal to the developing wall, the frontal epithelium is invaginating as the polypide rudiment. Production of a polypide in cheilostomes seems to follow the initiation of a cross wall.

The zooid bud in *Membranipora* is very long ('giant bud'), and the transverse walls clearly arise after the basal wall has been laid down. When the bud is shorter, not or scarcely exceeding the length of a zooid, the transverse wall may form terminally, as in *Metrarabdotos*,[5] so that it is laid down as a continuation of the basal wall; or the transverse wall may arise by subsequent invagination as in *Membranipora*. In other cheilostomes, the distal end of the nearly formed zooid may calcify except for a membranous central area, which subsequently swells out as the next zooid bud (as in the right-hand zooid depicted in Fig.5A, p. 39, and in Fig.12H, p. 89).[47,89] This situation is further discussed in Chapter 5.

In *Bowerbankia* and its allies stolonal elongation derives from apical growing points, behind which perforate septa are produced at intervals so forming a linear series of kenozooids.[7] Mesenchyme cells from the growing tip aggregate in strands which unite to form the stolonal funiculus. Septa arise by annular invagination of the ectoderm, when the terminal cystid has reached its appropriate length. The two cell layers of the septum secrete a cuticular lamella between them. One or more buds, often a series, appear on the surface of a newly delimited kenozooid, each bud containing a polypide rudiment and developing into an autozooid.

Calcification

The cuticle is composed predominantly of protein but also contains mucopolysaccharide, including a substance loosely described as chitin. Bryozoan chitin, sometimes at least, is a stereoisomer of true chitin (specifically the β-polymer of N-acetylglucosamine). The cuticle of a frontal membrane or the superficial layer of a wall has the form of a hyaline pellicle; deeper in the wall the cuticle is fibrillar and provides the matrix on which calcium carbonate crystals are deposited; finally, innermost and adjoining the epidermis, is a thin homogeneous layer.[28,30]

As in other biological systems (very little is known specifically for bryozoa) organic and inorganic materials must be secreted from the epidermis into a fluid-filled reaction space. Carbon dioxide and HCO_3^- in the body originate directly from the sea or by decarboxylation in the epidermis, as during the Krebs cycle. Carbonate may form from HCO_3^- in the reaction space either by the addition of hydroxyl ions

$$HCO_3^- + OH^- \rightleftharpoons CO_3^{--} + H_2O$$

or through the removal of CO_2

$$2HCO_3^- \rightleftharpoons CO_3^{--} + CO_2 + H_2O.$$

Calcium originates directly or indirectly from the sea, and the ions are transported across the epithelium. In *Electra*, Bobin and Prenant[3] were able to demonstrate (by von Kossa's silver method) a build up of calcium salts in the cuticular matrix before any crystallization could be detected.

$CaCO_3$ is deposited on the matrix in one or other of the crystal forms calcite and aragonite (rarely both): the circumstances deciding which are not understood. It has been established that calcitic skeletons contain very little strontium ($<0{\cdot}6\%$ $SrCO_3$) but substantial amounts of magnesium ($>7{\cdot}4\%$ $MgCO_3$), whereas aragonitic skeletons are richer in strontium (about $1{\cdot}4\%$ $SrCO_3$)—and could therefore have practical value as a means of monitoring ^{90}Sr in sea water—but poor in magnesium (about $0{\cdot}6\%$ $MgCO_3$).[82] This does not necessarily indicate that concentrations of Mg^{++} and Sr^{++} determine crystal form; indeed both induce aragonite formation when $CaCO_3$ is precipitated from a solution of $Ca(HCO_3)_2$ *in vitro*. It is not improbable that the organic nature (e.g., amino acid composition) of the cuticle, or dissolved organic materials in the reaction space, influence crystal form.

Primary calcification in cheilostomes usually takes the form of calcite. The manner of deposition may differ between the ancestrula and the subsequent zooids. In the former it must be added to an otherwise fully formed wall, and has the nature of a thin layer of crystal spherites secreted below the hyaline pellicle. Deposition, as seen in *Escharoides* (Gymnocystidea), proceeds from all around the periphery of the frontal wall towards the orifice.[59]

In subsequently formed zooids the calcite can be incorporated as crystal spherites[30,80] within the wall as it develops behind the growing edge of the colony. Later, in *Bugula*,[80] the calcite forms crystal threads exhibiting strong birefringence under polarized light. Calcification starts in the proximal—and therefore first formed—part of the cystid and spreads distally until halted at the site of the frontal membrane (in Anasca)[30,80] or orifice (in Gymnocystidea and Ascophora).[59] Deposition continues around the orifice of the zooid, and the primary calcified wall is complete.

Degeneration and regeneration of polypides

There is no excretory system as such, though coelomocytes in the body cavity apparently absorb waste materials. Insoluble products may be stored or discarded through degeneration of the polypide, forming what is known from its appearance as a *brown body* (Fig.5E, p. 39); but excretion cannot be regarded as the prime function of this phenomenon.

Examination of the colonies in many bryozoa reveals the presence of one or more dark spheroids in older zooids. These are formed by degeneration of the polypide, as may happen several times during the life of the zooid. Much of the softer tissue disintegrates and is phagocytized; but the midgut region, already stained brown with cytoplasmic inclusions, persists as an undestroyed rounded mass. Meanwhile a new polypide forms in the usual manner, though from the distal wall of the cystid. In some gymnolaemates the future digestive tract of the replacement polypide forms around the brown body, so that the latter is ultimately voided from the anus; in other bryozoans, including cyclostomes, the brown bodies accumulate in the coelom, loosely bound by mesenchyme, until several are present. In stoloniferous ctenostomes, e.g. *Bowerbankia*, *Walkeria*, *Zoobotryon*, the cystid may drop off the stolon after a brown body has formed; the stolon then produces a new zooid bud.

Cyclostome zooids can, under certain circumstances, be closed by a calcareous diaphragm which is dissolved when the new polypide forms below it.

Polypide degeneration certainly occurs in response to unfavourable conditions, and is a predictable response to placing freshly collected colonies in indifferent aquarium conditions; it may take place at the onset of winter; it also results frequently from associated reproductive activity, such as brooding of embryos. At other times it is perhaps a natural consequence of senescence in the polypide, as these appear to live for a few weeks only.

COORDINATION

Each zooid contains a small central nervous system and sometimes a dermal nerve network. The ganglion is dorsal, lying below the lophophore on the anal side of the pharynx. It is situated within the mesocoel or ring coelom, and partly blocks the opening between the two parts of the coelom. The ganglion is the best developed part of a circumpharyngeal nerve ring, from which one pair of sensory fibres and one pair of motor fibres pass into each tentacle. Other fibres pass to the tentacle sheath and alimentary canal.

A question that awaits solution is whether or not all bryozoans possess a colonial nervous system. The presence of fibres in the funiculus passing from zooid to zooid has been both reported and denied in the past, but a recent anatomical study by Lutaud[60A] has shown conclusively that such fibres are present in *Electra*. She describes in each zooid a single fibre, which passes all around the basal border of the interzooidal walls and connects through the mural pores with the similar peripheral fibre in each of the adjacent zooids. Two fibres following the tentacle sheath link the peripheral fibre with the main ganglion. As Marcus[63] established that no zooid responds to a stimulus applied to a neighbour, the physiology and role of this colonial nervous system obviously require investigation.

It is similarly obscure what colonial function the funiculus of gymnolaemates has, unless the transmission of metabolites proceeds more easily along mesenchymal strands than it would through open or plugged pores. There is no evidence for the coordination of zooids by means of the funiculus.

There are touch sensitive cells in the tentacles and perhaps a sense organ in the avicularium (p. 92). Larvae often have 'eye-spots', but a photoreceptive function has never been demonstrated experimentally;[17] moreover, larvae without 'eye-spots' can respond to directional illumination.[73] The branch apices in *Bugula* are positively phototropic, but the physiology of photosensitivity is not understood.[80]

In the coordination of ovicell development (Chapter 5) there is evidence for the formation of a hormone; but there is none yet for their production within the colony as a whole, although the existence of physiological gradients might be interpreted as suggesting their presence.

Physiological gradients

It is clearly established that physiological gradients exist in a bryozoan colony. This may sometimes be revealed by the distribution of reproductive zooids and degenerating polypides, such that zones respectively of developing zooids, ovicellate zooids, zooids with embryos present and polypides degenerating, zooids with mature larvae and brown bodies, and zooids with empty ooecia and regenerating polypides may be present; and the sequence may be repeated until the senescent zooids in the oldest part of the colony are reached.

Powers of growth and regeneration are greatest near the growing edge or growing points, and decline proportionally to the distance from the budding zone. There is evidence of polarity in *Zoobotryon* where cut lengths of stolon from the youngest part of the colony regenerate only at their distal end, whereas pieces of older stolon regenerate from both ends. That the protein content of the coelomic fluid falls *pari passu* with declining regenerative ability[33] appears to be a correlative of the gradient rather than its cause. It is plain that much remains to be learned about the individuality of the colony and the mechanisms of its internal control systems.

4

ECOLOGY—
THE GYMNOLAEMATA AND THEIR ENVIRONMENT

BATHYMETRIC DISTRIBUTION

The shore

Bryozoans are found from between tidemarks down into the great ocean deeps (a cellularioid was taken in the Kermadec Trench by the Danish 'Galathea' Expedition at 8210–8300 m). Few are primarily intertidal. On British shores *Alcyonidium hirsutum*, *A. polyoum*, *Bowerbankia imbricata*, *Electra pilosa* and *Flustrellidra hispida* (Fig.1A), all associated with midlittoral fucoids, and *Cauloramphus spiniferum*, *Cribrilina cryptooecium* (Fig.5C) and *Cryptosula pallasiana* on stones, are among the most characteristic species. The underside of boulders and cut-away cliff faces also provide suitable niches, with *Cryptosula pallasiana*, *Schizomavella linearis*, *Schizoporella unicornis* (Fig.5I) and *Umbonula littoralis* incrusting the rock, together with the bushy growths of *Scrupocellaria* and, in summer, the pendent colonies of *Bugula* (Fig.1B,F).

Where soft rock has been eroded by the sea into deep recesses, as at Tenby in South Wales, only a few small, shade-tolerant algae compete with the carpet of sedentary animals such as anemones, ascidians, hydroids, and so on, which flourish in the absence of smothering seaweeds on the continuously moist surface. These communities contain a wealth of bryozoan species, many of which are more frequently encountered in the permanently submerged environment beyond low tidemark. The kelp fringe, through which the infralittoral region merges with the shore, also supports a rich associated fauna. *Membranipora membranacea* covers the laminarian fronds, while *Celleporina hassallii*, *Escharoides coccineus* (Fig.5F), *Umbonula littoralis* and others incrust the holdfasts.

The infralittoral zone

The greatest abundance of bryozoans occurs in the shallow waters of the continental shelf, which usually terminate at a depth of 200 m or a little more. It appears, however, that the vertical distribution of bryozoans is far from uniform in shelf waters, at least in the Mediterranean for which the best data are available.[45] The greatest diversity of species and their maximum abundance lie between 20 and 80 m, with a peak of 40 m (Fig.8). Many bryozoans are thus fairly stenobathic and restricted to the well-lighted or infralittoral zone, although not themselves directly dependent on illumination.

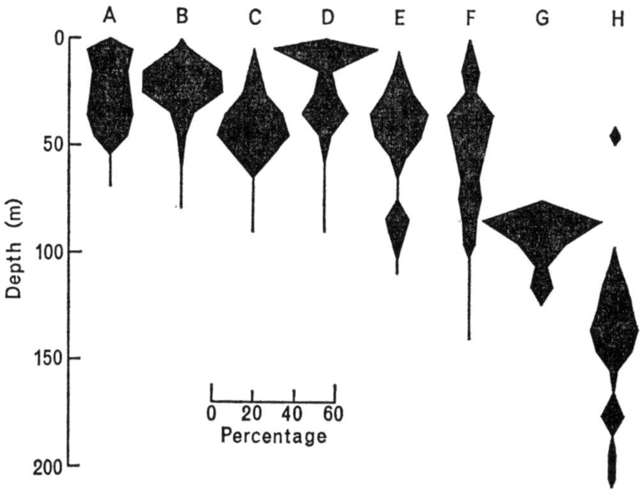

Fig. 8 Patterns of depth distribution

'Kite diagrams' showing the bathymetric range of some western Mediterranean infralittoral and circumlittoral cheilostomes

A *Margaretta cerioides*
B *Hippopodinella* spp
C *Schizoporella magnifica (Myriapora truncata* is similar)
D *Savignyella lafontii*
E *Schizomavella discoidea* (*Fenestrulina malusii* is similar)
F *Cellaria salicornioides*
G *Ramphonotus minax*
H *Escharina dutertrei* (occasional colonies occur at greater depths)
The bimodal distributions of D and E reflect the availability of suitable substrata.

(Data from Gautier[45])

The circumlittoral zone

Beyond the infralittoral zone Pérès[158] recognizes a circumlittoral zone characterized by sciaphilous (shade-loving) algae. One Mediterranean community in this zone is dominated by the kelp *Laminaria rodriguezii*, with which *Haplopoma impressum* is particularly associated; another facies is characterized by the red alga *Halarachnion* bearing such bryozoan epiphytes as *Aetea*, *Nolella* and *Walkeria*.

The richest circumlittoral Mediterranean community, however, is the coralligenous biocoenosis in which coralline and soft dwarf algae, serpulid worms, gorgonians, various corals and many types of bryozoan abound. The latter include *Adeonella calveti*, *Frondipora reticulata*, *Myriapora truncata*, *Pentapora foliacea*, *Sertella* spp. and *Smittina cervicornis*, all of which are erect branching forms with rather brittle colonies (Fig.1H,I,J). The great abundance of such forms reflects the absence both of large algae and of violent water movement.

The deep sea

On the continental slope and beyond, rocky outcrops bear characteristic assemblages of bryozoans together with brachiopods, corals and so on. Schopf[1] has noted that the maximum number of species per dredge station, based on all the available data, decreases from as many as 64 in 300 m, to 15 in 1000 m and 5 below 2000 m. Beyond the slope, absence of hard substrata probably restricts the occurrence of sessile epifauna. Here bryozoans are generally of the erect type and cellularioids, such as *Levinsenella* and *Himantozoum*, predominate. They form bushy colonies attached to the shells of foraminiferans by a stalk of kenozooids.

Another characteristic feature of abyssal bryozoans appears to be that the tentacles are relatively longer than in comparable shelf species and can be protruded a very long way. Stomach contents from material preserved shortly after capture (Schopf[1]) included detritus but no recognizable organisms, suggesting that the zooids filter detritus in suspension over the bottom.

ENVIRONMENTAL FACTORS IN BATHYMETRIC DISTRIBUTION

Temperature

In the west Norwegian fjords the vertical distribution of bryozoans indicates a bathymetric boundary lying somewhere between 40 and 60 m. twenty-six species were recorded solely shallower than this, 45 exclusively deeper, and only 9 were common above and below.[75] Here the distribution of bryozoans apparently correlates with the hydrography, for the widely fluctuating temperatures (<3 to >15°C) and salinities (<29 to $34^0/_{00}$) of the surface layer are stabilized within narrow limits (6–10°C; about $34^0/_{00}$) at 50 m.[23]

A physiological interpretation of depth distribution in relation to temperature can only be speculated upon at present. One striking hydrographic feature paralleling the bathymetric distribution of Norwegian bryozoans is that depths of less than 40 m are characterized by a marked rise of temperature in summer. Shallow water species perhaps require a high summer temperature (Fig.10, p. 84) for the initiation of reproduction, whereas species living at greater depths find increased temperatures unnecessary or harmful. The restriction of so many Norwegian species to depths greater than 40–60 m, and the apparent absence of such a group in lower latitudes,[23] suggests inability of such species to withstand the severe cold experienced near the surface during the boreal winter.

Species which are eurybathic appear, from their world distributions, to be eurythermic.

Wave action

The effects of wave action on vertical distribution are far from clear. Water movement appears essential for some species at least, although too much turbulence must be damaging. The severity of wave action to which a coast is subjected depends to a large extent on the proximity of sheltering land masses, for the size of waves at any time depends on the following factors: the velocity of the wind, the distance it blows across the water (fetch), the depth of this water and, to a smaller extent, the length of time the wind has been blowing. As waves enter shoaling water their character becomes affected by the configuration of the sea bed. Interaction between waves and the bottom commences at least by the time water depth d is equal to $0{\cdot}5L$, where L is the wavelength or distance between successive crests. For a big ocean swell such interaction may commence at the very edge of the continental shelf ($L=400$ m; $d=0{\cdot}5L=200$ m). On exposed European coasts winter storm waves of length 60–100 m are usual, so that d falls between 30 and 50 m.

The zone of swash created by the distortion of orbits as waves approach the shore and break upon it is related to wave height H, with the lower limit at about $2{\cdot}5H$.[158] Storm waves of the length just mentioned would be 4–8 m high, so that the limit of multidirectional turbulence would lie between 10 and 20 m.

While there may be no direct connexion, it does at least seem noteworthy that in the Mediterranean the commonest type of infralittoral distribution among bryozoans (Fig.8C) displays rapid attenuation above 20 m and below 50 m.[23,45] Thus wave action may be one of the factors controlling vertical distribution in shallow water, especially affecting the upper limit. There remains, however, a general shortage of quantitative data on the bathymetric distribution of bryozoans and other sedentary animals, which could be obtained most readily by free-diving techniques, and also of measurements of the essential environmental parameters which might affect their distribution.

From the limited data that are available, it is apparent that, if wave action is a controlling factor around British coasts, substantial differences in depth distribution would be expected between exposed and sheltered localities. Thus, at a depth of 30 m on a moderately exposed Irish Sea coast a particle speed of 10 cm/sec will be exceeded on about 25 days/year; while on an exposed Atlantic coast 10 cm/sec will be exceeded on about 300 days/year.[154] For a particle speed of 30 cm/sec the corresponding figures are 4 and 90 days/year, and for 50 cm/sec they are <1 and 30 days/year. Expressed another way, the spectrum of particle speeds characterizing a given depth off-shore in the Irish Sea will be found (very roughly) at three to four times that depth off an Atlantic coast.

Availability of substrata

Apart from temperature and wave action, many of the established bathymetric patterns may relate primarily to the availability of suitable substrata, for many bryozoans seem particular in respect of their chosen support. On a rocky bottom in the western Mediterranean the alga *Cystoseira* dominates an important biocoenosis and is host to many bryozoan epiphytes, notably *Schizobrachiella sanguinea* which forms cuff-like incrustations around the base of the plant. On soft substrata the endemic Mediterranean eel-grass *Posidonia oceanica* dominates large areas. *Electra posidoniae* and *Fenestrulina joannae* are virtually restricted to the foliage of this plant. The infralittoral zone is characterized by the presence of photophilous plants, and the disappearance of many species of bryozoan below 50–60 m may be related to the absence of these plants and thus, in the first instance, to the penetration of light.

COLONY FORM IN RELATION TO BATHYMETRIC DISTRIBUTION

Stach,[94] in a paper much quoted by palaeoecologists, has correlated colony form in bryozoa with their preferred depth range. It is not clear whether Stach based his conclusions mainly on *a priori* argument, for no ecological observations were published. Thus, he surmised, an incrusting habit was an adaptation to life in the littoral and infralittoral, or wave affected zones; as were the flexible foliaceous (*Flustra*) and bushy, jointed-stemmed colony forms (*Cellaria, Crisia, Margaretta*; Fig.1C,G). The brittle foliaceous (*Adeona, Pentapora, Smittina cervicornis*) and erect branching (*Myriapora, Hornera*; Fig.1H,J) growth forms would be found deeper, as in the circumlittoral community described above.

Stach placed the reticulate retepore colonies (*Sertella*; Fig.1I) in a category of their own, and regarded them as tolerating strong wave

Colony form	Rocky shore, south-west Wales	25–30 m, hard substratum with shells, Sound of Mull, Scotland	60–80 m, muddy to sandy, some shells, northern Gulf of Guinea, 'Atlantide' Expedition[40]	60–80 m, coralligenous biocoenosis, western Mediterranean (Cheilostomata only)[45]	90–140 m, stone-coral (*Lophelia*) reef, western Norway[75]	2000 m, world's oceans, 'Challenger' Expedition (Cheilostomata only)
Incrusting and nodular	58	71	48	63	60	9
Stoloniferous	5	6	4	0	0	0
Tufted	21	15	4	1	6	18
Arborescent	0	0	0	0	0	48*
Foliaceous (flexible)	3	3	0	1	4	3
Variously erect (brittle)	0	3	16	16	18	18
Reteporine	0	0	8	3	2	3
Erect (jointed)	13	3	4	10	8	6
Free-living (lunulitiform)	0	0	16	0	0	0

Percentage of species exhibiting each colony form in six habitats. Each sample is based on at least 25 species. (* This figure may actually be higher, for it is not always clear from the 'Challenger' Expedition report whether a long stalk was present or not.)

action. Few of the European species are found in very shallow water; although, in general, retepores are shelf forms. The species collected by H.M.S. 'Challenger' ranged from depths of a few metres down to 1000 m, but the 'Siboga' Expedition to the Indo-Pacific found them most commonly at less than 50 m, especially on hard *Lithothamnion* bottoms.

The colony forms represented in six faunal assemblages are analysed in the accompanying table. Stach's conclusions (and certainly the application to palaeoecology) are supported in general, but the incrusting type clearly predominates on all hard substrata, not just in the shallows and on the shore. The prevailing form in the deep ocean is tufted, usually with a stalk (arborescent). The free-living type is discussed in the following sections.

OTHER FACTORS AFFECTING DISTRIBUTION

Sand and mud

Bryozoa are usually found attached to a firm substratum, though a few exceptions occur and are discussed in this section. One of the exceptions is the cellularioid genus *Kinetoskias*, adapted for life in mud and recorded for example at the bottom of Norwegian fjords. The colony form is arborescent, with a tuft of branches arising from the top of a long stalk of fused kenozooids. The lower end of the stalk is anchored in the mud by a cluster of rhizoids.

Unique in many ways is the ctenostomatous genus *Monobryozoon*. The 'colony' here consists of a single autozooid of the *Nolella* type, from the base of which arise anchoring stolons (p. 31). One species, *M. ambulans*, lives in shell gravel and may be described as belonging to the interstitial fauna, a remarkable community of miniature forms living between the sand particles, to which nearly all the major animal groups contribute. A second species lives under more muddy conditions.[43]

Very different are the so-called lunulitiform (after the fossil genus *Lunulites*) bryozoa confined to warmer waters. A well-known example is *Cupuladria canariensis* which is distributed only within the limits of the 14°C isocryme. The colony is discoidal, conical or like a limpet shell (Fig.1D), up to 1–2 cm diameter, and normally not anchored to the substratum in any way. Autozooids and vibracula, the latter being heterozooids bearing well-developed setae (p. 95), are confined to the convex surface. The colonies may be found on sand, silt or mud, and can constitute a major part of the epifauna. Thus, in two widely separated parts of the Atlantic, the abundance of colonies has been estimated at 2000–3000/m^2. Such high numbers may perhaps be related to a method of propagation described in *Discoporella*[65] in which daughter colonies are budded off around the periphery of the parent.

The colonies live with the broad base of the cone downward[48,49]—the only orientation in which they would be stable to water movement—but may be lifted off the substratum by the stilt-like abducted setae of the peripheral and subperipheral vibracula. Since setae of the two series naturally make different angles with the substratum and readily adjust for irregularities of the surface, the colony possesses considerable stability. Nevertheless, lunulitiform bryozoans do not live in areas of turbulence, and are absent from the shore and immediate sublittoral. Their ability to survive in areas of silt deposition is described below.

Sediment deposition

Where a major river is discharging silt-laden water into the sea, there will occur beyond its mouth or delta a transition zone (described as fluviomarine) characterized by low surface salinity and a high rate of sediment deposition. The latter is the more harmful to benthic animals like bryozoans. Off the Mississippi delta, for example, the only species that can tolerate even moderate rates of deposition are those with lunulitiform colonies, such as *Cupuladria canariensis*.[56] The reasons for their success here are: (1) their independence of the substratum; (2) the power of the vibracular setae to sweep particles off the surface of the colony;[38] and (3) the ability of colonies to regain the surface should they become immersed.[38]

Within the European area, Lagaaij and Gautier[57] carried out an analysis of the bryozoan fauna in relation to deposition rates off the delta of the Rhône. This part of the Mediterranean is insufficiently warm to support a lunulitiform population. The study is of interest on account of the method employed, quite apart from the intrinsic value of the results.

Since the colony in any species of calcareous bryozoon is built up by replication of its own characteristic motif—the autozooid—in which the features necessary for identification are usually present, comminution after death produces fragments that are still readily recognizable by a specialist. These tiny pieces accumulate in sand and silt where, from a small sample, they can provide a record of those species living in the area surrounding each particular sampling point. Thus, when Lagaaij and Gautier examined 193 sand samples corresponding to a grid of stations lying offshore from the Rhône, they picked out 23,579 fragments belonging to 105 species!

The analysis not surprisingly revealed an almost complete absence of bryozoan fragments in samples from the fluviomarine silt fan off the Grand Rhône distributary, and it was considered most unlikely that this paucity arose simply from the diluting effect of the rain of clayey particles on organic remains. Away from the fan, but still within the area of slow silt deposition, fragments of one species only, *Cellaria fistulosa*, occurred in abundance, together with some *Scrupocellaria* and *Crisia*. Rather strikingly all three genera possess colonies of the erect, jointed type, which thus appears to be the best adapted habit—apart from lunulitiform—to cope with sediment deposition (Fig.1c). The loose clumps of *Cellaria* attach to a silty bottom by means of basal rootlets; they lack any extensive horizontal surfaces on which smothering particles could accumulate; and they can, presumably, grow upwards away from a gradually deepening deposit. Much the same may also be true of *Scrupocellaria* colonies which, like those of lunulitiform bryozoans, possess vibracula. Small numbers of species having a different kind of colony form, notably that with erect but brittle branches, appear at the periphery of the area.

The Mediterranean study provides clear factual evidence of the absence of bryozoans from areas of rapid silt accumulation. From the palaeoecological viewpoint it suggests that a facies in which jointed-stemmed bryozoans predominate is indicative of a fluviomarine environment. A core sample from beside the mouth of the Grand Rhône exemplified this. At the lowest level the rich fauna of the coralligenous biocoenosis was discovered. This gave way to a layer of cellariiform fragments as silt deposition accelerated until, eventually, in the uppermost layers, there were no bryozoan remains.

It seems appropriate to mention here another development, of particular interest to sedimentologists, arising from the technique just discussed. Work on sands along the coast of Holland[156] revealed the presence of fragments belonging to species known to be restricted to warm water. It became apparent that these originated by the erosion of fossiliferous outcrops of Eocene and Pliocene age situated some way to the south. The bryozoan particles were thus functioning as 'tracers' in a clear demonstration of north-going sand transport along the Dutch coast.

Salinity

The Gymnolaemata as a group are fairly stenohaline and restricted to normal sea water (salinity roughly $35^0/_{00}$). A few species are known from hypersaline waters. Thus the Cambridge Expedition to the Suez Canal[52] in 1924 recorded 22 species of bryozoa at Port

Taufiq and in the southern part of the Canal (roughly $44^0/_{00}$), five of which—including *Bugula neritina* and *Watersipora subovoidea*—had reached the central region of the Canal (about $49^0/_{00}$).

Reduced salinities result in an impoverished bryozoan fauna. This is clearly demonstrated by the decline in specific diversity between the Skagerrak ($>30^0/_{00}$; about 130 species) and the inner gulfs of the Baltic Sea ($<6^0/_{00}$; 2 species). At intermediate localities, 33 species have been recorded from Kiel Bay ($10–15^0/_{00}$) and 4 from the vicinity of Kursk Spit in the south-eastern Baltic ($7–8^0/_{00}$). One euryhaline species, *Electra crustulenta*, ranges from Kiel Bay right into the Gulfs of Bothnia ($<2^0/_{00}$) and Finland; another, the ctenostome *Victorella pavida* (so called from its discovery in the Victoria Docks, London) occurs in certain river mouths. Although salinity appears to be the major environmental component excluding bryozoans from the Baltic, the extensive occurrence of ice during winter constitutes a second unfavourable factor.

A similar attentuation occurs between the species-rich Mediterranean ($37–39^0/_{00}$), the Black Sea ($17–23^0/_{00}$; 19 species) and the Caspian Sea ($12–13^0/_{00}$; 4 species). In the Black Sea there is a complicating factor, however: the lack of oxygen and presence of hydrogen sulphide at depths beyond 150 m. The Caspian fauna includes an endemic form of *Bowerbankia* and two typically brackish species, *Victorella pavida* and the anascan *Conopeum seurati*. *Cryptosula pallasiana* appears to be the most euryhaline ascophoran, surviving even in the Sea of Azov ($12^0/_{00}$).

Around the British Isles, two marine species are notably euryhaline: *Conopeum reticulum* and *Bowerbankia gracilis*, which may be found together in salinities lower than $20^0/_{00}$. Truly brackish species are *Aspidelectra melolontha*, a membraniporellid from estuaries in eastern England, *Conopeum seurati*, *Electra crustulenta* and *Victorella pavida*. The three anascans (*Conopeum* spp. and *E. crustulenta*) have been so confused with each other that assessment of their distributions with respect to salinity is impossible, except that *C. seurati* appears to prefer enclosed waters and can tolerate salinities as low as $1^0/_{00}$.

The genus *Victorella* is distributed through marine, brackish and freshwater habitats. *V. pavida* is a true brackish water species of wide distribution,[32] occurring in a variety of biotopes, but often associated with reeds (*Phragmites communis*). In temperate latitudes *V. pavida* adapts to adverse winter conditions by producing resting zooids or *hibernacula*, the formation of which depends in a rather complex way on the interrelationship between temperature and salinity. It appears, from work in a brackish lake near Naples,[37] that winter temperatures do not result in the formation of hibernacula unless the salinity is less than 5‰. Yet the colonies rejuvenate in spring with a rising temperature (at about 15°C), while salinities may still be as low as 1‰. In some parts of this lake, *Victorella* survives summer salinities in excess of 35‰, but may produce 'summer hibernacula' in response to pollution or other unfavourable conditions. It seems possible that physiological races evolve, each having somewhat different tolerances.

At an estuarine locality in southern India, in which the salinity varies from less than 1‰ during the monsoon to 34‰ during the dry season, the period of abundant growth and reproduction of *V. pavida* is confined to the wet season when salinities are lower than 15‰. Colonies tested in the laboratory for 24 hours withstood perfectly fresh water and showed mortalities in salinities above 10‰. Nearer to the dry season, when ambient salinity was 14–22‰, zero mortality in experiments was obtained only in the range 16–23‰, and fresh water was lethal.[67]

Another ctenostome, *Paludicella articulata*, has become an entirely freshwater species. It, too, produces hibernacula in response to unfavourable conditions.

Water currents

Some possible effects of wave-generated water movements on bathymetric distribution have already been mentioned. Water currents, which lack the destructive element associated with waves, are favourable, possibly essential, for the success of many filter feeding organisms. Observations on the bottom fauna off the Isle of Man, in a region where surface tidal currents reach up to 4½ knots (2·3 m/sec) on spring tides, indicate in a general way the dependence of sublittoral bryozoans on water movement.[42] The current velocities close to the sea bed are unknown, but the nature of the substratum reflects their strength. As water movement declines so coarse sand, with gravel and stones, gives way first to fine sand and then to muddy sand. True mud occurs off the Isle of Man only beyond the 70 m isobath where both wave and current action are absent.

Maximum diversity of bryozoans (about 60 species) occurs on the coarsest ground. However, it is also on this kind of bottom that suitable support for the colonies of incrusting forms is most abundant. To allow for this Eggleston[42] calculated the number of colonies present per 1/10 m^2 of available support: over 500 on the coarse substratum, 50–100 on sand and <50 on the mud. It is possible that settlement will occur more readily, especially for species producing few larvae, where suitable support is more plentiful, but the figures apparently provide some indication of the value of water movement to bryozoans.

The effect of water currents is more obvious in very shallow water where characteristic assemblages of sponges, ascidians, serpulid worms and bryozoans occur in areas benefiting from particularly high mass transport of water. Such habitats may be large channels, like some Norwegian fjords or the Menai Strait in North Wales; narrows connecting almost land-locked bays with the sea, as at Lough Ine in Ireland; or as small as power station intake conduits or ships' cooling pipes.

In the narrow passage (rapids) at the mouth of almost enclosed bays on the west coasts of Ireland and Scotland, in which high water movement is combined with the absence of wave action, the fronds of laminarian algae may be smothered with bryozoan colonies. Four of the commonest species are *Callopora lineata* (Fig.5A), *Hippothoa hyalina* (Fig.5G) and *Microporella ciliata*—all incrusting—and the bushy, rhizoid-attached *Scrupocellaria reptans*. In western Norway *Callopora craticula* and *Cribrilina annulata* would be added to the list.

The most thorough study of a rapids system so far concerns Lough Ine in southern Ireland.[155] Water flows in and out from the open sea through a channel which narrows to a minimum width of 12 m; here there is a sill only ½ m deep at the lowest tides. Water flow is fastest at this point, and a maximum velocity of 3 m/sec is attained. Except at the sill, where *Laminaria digitata* prevails, the dominant kelp is *Saccorhiza polyschides*. In general the bryozoan species found in the turbulent part of the Lough Ine rapids are those characteristic of open shores subject to wave action: *Alcyonidium hirsutum*, *Hippothoa hyalina*, *Celleporina hassallii*, *Electra pilosa*, *Escharoides coccineus* (Fig.5F), *Scrupocellaria reptans* and *Umbonula littoralis*.

In the moderately fast flowing water towards each end of the narrows a luxuriant growth of *Membranipora membranacea* occurs on the kelp fronds; but this species is absent from the vicinity of the sill. There is negligible development of this bryozoan in regions where the maximum water velocity is less than 1 m/sec, and it reaches peak abundance in those areas experiencing maximum velocities of 1·0–1·5 m/sec, there during September covering 30–40% of the frond area. Above 1·6 m/sec there is an abrupt decline to a cover of 1%.

Hippothoa hyalina prefers less vigorous flow conditions. Where the maximum velocity is in the range of 0·3–0·6 m/sec, the *Saccorhiza* carries 100–500 colonies per frond, with perhaps as many again on the frills of the stipe. Somewhere between 0·6 and 1·0 m/sec *Hippothoa* ceases to occur on the fronds but remains abundant on the frills where, presumably, water movement is less strong. Where the influence of the open sea becomes marked, the *Hippothoa* vanishes even from the frills and is confined to the holdfast itself.

A laminarian canopy demonstrably retards the current flow, just as it buffers wave action. What probably matters to a bryozoon, however, is the amount of shear in the boundary layer between the attachment surface and the flowing water: not the velocity of the main flow, although that may be the parameter most easily measured. At the actual alga-water interface there is no relative movement between phases at all: the water velocity near the frond thus approaches zero, however great the water current may be a short distance away.

The velocity gradient could affect feeding, but its main impact would be on larvae ready to settle, as Professor D. J. Crisp has shown for barnacle cyprids. While the distribution of the *Hippothoa* colonies in the Lough Ine rapids probably reflects optimum settlement conditions, the cover of *Membranipora* may represent a balance between conditions facilitating the settlement of prospecting larvae and those—in terms of food and oxygen supply—favouring prolific somatic growth. It is clear that these field observations require following up with laboratory experiments.

MARINE FOULING

Any surface immersed in the sea becomes covered with sedentary organisms. When that surface is a ship's hull or some other man-made structure, whose performance will be affected by such growths, the settlement that results is termed 'fouling'. Fouling affects not only ships but any industrial plant—such as a power station or refinery—using sea water for cooling purposes. Intake conduits must periodically be cleaned out to maintain a clear channel. Ships' cooling pipes may also foul badly unless made of copper, which is toxic to settling larvae, or a cupro-nickel alloy, which is more resistant to corrosion than copper alone.

The passage of a ship through water is hindered by resistance depending on (1) the displacement of water by the hull and (2) the frictional forces between the water and the submerged surface.[160] Only the second of these is affected by fouling. The frictional resistance R_f of a surface moving through water may be expressed by the empirical formula

$$R_f = fSV^n,$$

where the value of f, the coefficient of frictional resistance, is dependent on the character of the surface; S is the area of wetted surface; V is the velocity; and n is approximately equal to 2.

It has been shown that the value of f may increase from about 0·01 for a newly painted surface to 0·02 with moderate fouling and 0·03 with heavy growth. In the same way, n may increase from 1·9 through 2·0 to 2·1. Thus fouling can more than double the value of R_f as many tons of growth accumulate on the hull.

In practical terms fouling has meant that, until recent times, one year or even less at sea after repainting could result in a doubling of total hull resistance or, expressed another way, in a doubling of the propeller shaft horsepower needed to maintain the vessel's performance. It was considered that the hull resistance increased on average by ¼%/day in temperate waters, which resulted in a 50% increase in fuel utilization in 6 months. In the tropics resistance increased at ½%/day. The cost, in terms of increased fuel consumption and frequent dry docking for cleaning and painting ships' hulls, was clearly enormous. Copper-based antifouling paints are used to combat settlement, and recent improvements in paint technology now permit a period of two years between one dry docking and the next. Nevertheless, fouling still represents an economic factor of huge proportions.

Although over 50 species of bryozoa have been recorded on ships' hulls, in terms of weight they are less important fouling organisms than barnacles. Certain species, however,[76,78] notably *Bugula neritina*, *Schizoporella errata*, *Watersipora subovoidea* and *Zoobotryon verticillatum*, occur regularly and abundantly in warm waters. The tufted colonies of *Bugula* and *Zoobotryon* can create a great nuisance by clogging intake pipes of the cooling system. The special importance of *Watersipora* as a hull colonizer relates to the fact that it appears to be just about the most copper-resistant macrofouler known.[23] If it becomes established on recently painted surfaces, its extensive crusts provide a safe substratum for the attachment of barnacles, serpulid worms and branching bryozoans like *Bugula*, which together have such a marked affect on frictional resistance.[23,76]

A side result of ship fouling, of zoogeographical interest, relates to the spread of species beyond their original area of distribution.[23] It seems certain, for example, that *Watersipora* was introduced to Australian ports and from there, between 1956 and 1958, to Auckland, New Zealand. *Bugula neritina* has arrived in Britain, where it flourishes in certain south coast ports; while *B. flabellata* has been carried to Australia. Finally, the recent appearance of the brackish water species *Conopeum seurati* in the Caspian Sea suggests a ship-assisted passage through the Volga-Don canal. Dr Anna Hastings has noticed that this species produces dormant, closed zooids, corresponding to the hibernacula of *Paludicella* and *Victorella* discussed on p. 71, which would no doubt promote survival during short periods of wholly adverse conditions.

INTERRELATIONSHIPS

With algae

It has already been noted that bryozoans are not distributed haphazardly, but often occur in association with particular substrata. Examples have been given of species found on fucoids, laminarians, sublittoral algae and eel-grass; others prefer to grow on rock surfaces, shells, hydroids and other bryozoans. The most studied relationship is that existing between epiphytic bryozoans and algae in European waters.

There is substantial evidence that the settlement of bryozoans, as of other invertebrates, is far from an affair of chance: larvae search for, and metamorphose after finding, a suitable substratum. Newly released larvae of *Alcyonidium* and *Flustrellidra* have been offered a choice of algae on which to settle. The results established very clearly that the larvae settle preferentially on a few species of alga, and that such preferences tally very closely with the observed distribution of adult colonies on the shore (Fig.9).[72,74]

The only exceptional result in the experiments was that bryozoans normally living on *Fucus serratus* settled in moderate numbers on *F. spiralis*, which occupies a zone near the top of the shore and is devoid of epiphytes. Despite this, the results strongly suggest that algal selection—the choice of *Fucus serratus*, *Chondrus crispus* and *Gigartina stellata* by larvae of shore-living bryozoans—may be one of the mechanisms by which sessile organisms locate an appropriate tidal level. All three algae inhabit the lower shore.

Surface contour also affects settlement behaviour, for larvae select hollows or grooves in which to metamorphose, but is of secondary importance to the nature of the substratum. It has been demonstrated that *Fucus serratus* contains an extractable product which can be transferred to an inert substratum rendering it attractive to larvae.[41]

An even more familiar association exists between *Membranipora membranacea* and kelps such as *Laminaria digitata* and *L. hyperborea*. The *Membranipora* is structurally well adapted to life adhering to flexible, swaying fronds. Each lateral zooid wall contains a pair of breaks in the calcification, so that bending takes place without cracking the colony. Growth of *Membranipora* exhibits an interesting adaptation, such that the relationship between alga and bryozoon must be considered somewhat further.

Growth in *Laminaria* is achieved mainly by cell division in the transition zone between the stipe and the frond but, whereas the stipe is perennial and increases in length over a period of years, the frond

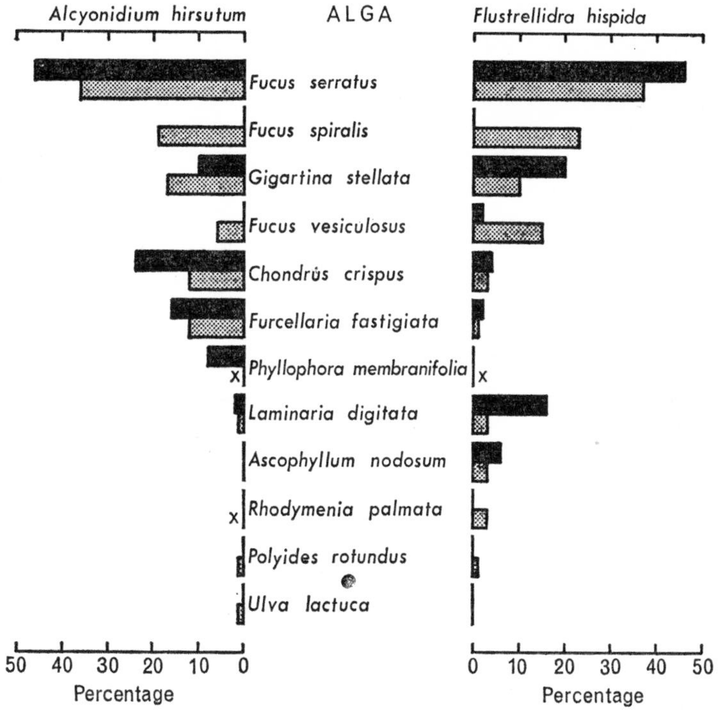

Fig. 9 Association of two bryozoans with algae

The percentage occurrence of *Alcyonidium hirsutum* and *Flustrellidra hispida* on various algae on some shores in Wales is shown in black; the percentage settlement obtained in experiments is shown in stipple. Where an algal species was not tested, X is shown.

is an annual production. The period of rapid frond growth is from January to May–June and later in the year only a narrow band of tissue is produced. This band joins that year's growth to the frond newly forming until, during winter, the old frond gets stripped off by wave action, or is finally cast in April–May.

Membranipora has one generation each year, with egg production occurring during spring and early summer. The larvae spend some weeks in the sea, growing and developing, and are most abundant in British coastal plankton between June and August. Settlement thus occurs on new but fully formed *Laminaria* fronds during summer and early autumn. Growth continues until the onset of winter, and colonies may reach a vast size (up to 2 million zooids).[148] Many colonies will be partly or wholly destroyed when the fronds disintegrate but, in spring, gametogenesis takes place and fresh growths arise from the overwintered stock.

The interesting feature of growth in *Membranipora* is that, instead of spreading evenly around its edge, the colony expands predominantly in a single direction. Thus a clear axis emerges and, to an almost exclusive degree, growth proceeds down the frond towards the stipe.[79] Ecologically the advantages of this are clear: the colonies grow onto the new frond from the old early in the year; later they approach the base of the frond, in which position they are most likely to survive the winter.

While it has been established that branch apices in *Bugula* are positively phototropic,[80] colonies of *Electra pilosa*, another bryozoon found on *Laminaria*, show no growth response to light.[63] Detached colonies of *Electra* may put out stolon-like outgrowths which are positively geotropic; but this is not a normal process. The mechanism underlying oriented colonial growth in *Membranipora* is at present obscure. Since *Laminaria* fronds when covered by water—which is most of the time—neither hang downwards nor float straight upwards, it may be supposed that light and gravity are not involved.

It appears possible that *Membranipora* can, in some way, detect the youngest part of an algal thallus and grow towards it. Settling bryozoan larvae can certainly do this. *Alcyonidium polyoum* larvae, when offered a choice of *Fucus* frond tips, the region just below the tips ('sub-tips'), and pieces from the centre and base of the frond, settled in the following manner:[72]

Tip	Sub-tip	Centre	Base
177	133	88	38

For colonies, rather than prospecting larvae, to detect an age gradient, appears to raise problems of perception and coordination about which we have no knowledge, and further research into this surprising phenomenon is obviously needed. An alternative possibility, however, is that the direction of growth is determined by water flow. Because *Laminaria* fronds always lead downstream in a current, growth by *Membranipora* towards the stipe could be a rheopositive response.[79]

With shells

Shells of living or dead bivalve molluscs provide a favourable substratum for many bryozoans. Scallop shells seem more suitable than those of mussels (*Modiolus*) and *Cyprina*. In an area off the Isle of Man[42] the occurrence of bryozoan colonies per unit area on the inner surface of bivalve shells was noted (expressed as a percentage of the number found on *Chlamys*):

Chlamys opercularis	100	*Modiolus modiolus*	55
Pecten maximus	83	*Cyprina islandica*	45
Glycimeris glycimeris	66	stones	32

The greater favourability of *Chlamys* and *Pecten* may relate to the ribbed nature of the shell, with the grooves attracting rugophilic larvae (cf. p. 76). On many shells bryozoans are more abundant on the inner surface, a circumstance perhaps related to contour or texture or both. It is however interesting to realize that, on a current-swept bottom, shells always lie in the stable orientation: concave side —and bryozoans—facing downwards. Protection of prospecting larvae from being swept away may be important in such an environment. In some species which form nodular colonies, the larvae apparently prefer a convex surface.[42]

A more specialized mode of life, which has evolved at least twice among present-day ctenostomes, is that of boring into the calcified structure of shells, serpulid tubes and barnacle plates. In the genera *Spathipora* and *Penetrantia* the slender stolons ramify in the substance of the shell, bearing at intervals zooids which gain access to the exterior through small perforations.

The bryozoans have no apparatus for mechanical tunnelling and it seems quite certain that the shell is dissolved with phosphoric acid. Silén,[90] following up the clue that easily detectable amounts of phosphorus were present in bored shells, but absent from unbored shells, applied a histochemical test to shell fragments containing living colonies of *Penetrantia*. Treatment with ammonium molybdate followed by the reducing agent amidol, which colours phosphate blue, revealed the presence of dense concentrations in the stolon tips. Phosphoric acid must be a very effective agent for boring in shells: a weak (0·0015 M) solution dissolved 3·6% by weight of a *Mytilus* shell in ten days, and strong solutions have a drastic effect. Little seems to be understood of the boring mechanism in the worm

Polydora or the sponge *Cliona* for example, despite recent work on the problem, so that comparisons outside the phylum are impossible.

A species of *Penetrantia* has just been described[93] which tunnels in the coenecium of the pterobranch *Cephalodiscus*. The mechanism of boring is not known.

Commensals

Members of the genera *Hippoporidra* and *Hippopodinella* habitually live on gastropod shells occupied by hermit crabs (Paguridae of several genera), exactly as does the better known hydroid *Hydractinia*. The bryozoan colony often extends freely beyond the shell opening, forming a cowl about the pagurid. The development may even be sufficiently extensive to trap the crab.[39] Hincks[146] reported that in some instances the *Hippoporidra* dissolves away the gastropod shell (although recent work has not confirmed this), just as the zoanthid *Epizoanthus incrustans* does in the same situation. An important advantage to the bryozoon is that it can survive successfully on a silty sea bed which otherwise lacks sites for attachment.

Crustaceans which have not moulted for some time frequently carry bryozoans. *Triticella*, however, is found only on the carapace or appendages, including the mouth parts, of decapods (e.g. *Nephrops*, *Calocaris*, *Goneplax*).

The most specialized, and not uncommon, bryozoan commensal is *Hypophorella*, a stoloniferous species which inhabits the papery lining layers of the tube of the polychaete *Chaetopterus*. Stripping off some of the lining and examining it under a microscope readily reveals the presence of *Hypophorella*. The stolons at first ramify over the inside of the tube, but they become embedded as the worm deposits additional lining layers. The zooids are well adapted to meet this contingency: each one is provided with a file-like lip to the orifice, with which it can rasp its way into the tube lumen. Well protected, the *Hypophorella* benefits from the feeding current created by the worm.

Bryozoans also act as host to commensals. Small entoprocts (*Loxosomella*) have been recorded on many species, where they aggregate near the orifices. The hydroid *Zanclea*[53] is also a frequent associate of bryozoans, occurring on retepores and other cheilostomatous forms, while *Halocoryne epizoica* lives on *Schizobrachiella sanguinea*.[45]

Another interesting example concerns a recently described barnacle (*Kochlorinopsis discoporellae*) commonly found associated with the lunulitiform colonies of *Cupuladria* and *Discoporella*. The cyprid bores into the upper surface of the bryozoon and then metamorphoses. A slit remains between the zooids through which the cirripede protrudes its cirri for feeding. The barnacle obtains a protected attachment site over an otherwise inhospitable sandy sea bed.

Predators

Bryozoans have been recorded, though perhaps taken incidentally, among the stomach contents of eider ducks (*Somateria*); and they are also eaten by fishes, notably wrasses (Labridae). The well-developed teeth of the latter are adapted for browsing on incrusting organisms, although bryozoans are likely to form only a small proportion of their diet.

More important as predators are sea urchins. The common *Echinus esculentus*, from the smallest size upwards, lives on and among the sublittoral laminarians. It does feed on the alga itself, but preferentially consumes the incrusting bryozoans and hydroids, both on the stipe and on the frond. Although echinoids are omnivorous, when specimens of *Strongylocentrotus* were given a choice of clean alga or *Membranipora*-covered alga, they much preferred the latter.[159] If an urchin browses over an area of 60–70 cm^2 per day, a great deal of *Membranipora* must be consumed. Another urchin, *Psammechinus miliaris*, has been noted scraping *Electra pilosa* off the shells of mussels (*Mytilus*). Echinoids also browse over rock surfaces and may thus incidentally feed on a number of incrusting bryozoan species.

Among molluscs, chitons have been noted preying on bryozoa in New Zealand.[47] Better known as specific predators are the nudibranchs. Many of these (Goniodoridae) are suctorial feeders with a narrow lancing radula which drills into the bryozoan zooid, and a pharyngeal pump for withdrawing the fluid contents. Examples are *Goniodoris*, which often feeds on the fleshy colonies of *Alcyonidium* and *Flustrellidra*; an *Onchodoris* which is a specialist feeder on *Cellaria*; and *Acanthodoris pilosa* which utilizes a wide range of species including ascophorans.[157]

Corambe pacifica attacks *Membranipora*. The smallest individuals enter the zooids and devour the polypide; larger ones attack the growing edge. More destructive are species of *Polycera* which, in various parts of the world, browse on *Membranipora*. Nudibranchs are voracious predators, consuming up to 30 or 40% of their own weight in food daily. *Polycera*, with its wide radula, ploughs through the zooids in any part of the colony, eating part or all of what it destroys.

Although the life history of *Polycera* has not been studied in detail, it is probably similar to that of *Adalaria*, which feeds on *Electra pilosa*, and has been studied by Dr T. E. Thompson. Spawning takes place in late winter and spring. The veligers seek living colonies of *E. pilosa* and will settle only in the presence of that species. The nudibranch is annual; it spawns once and then dies. *Polycera* likewise appears in spring, as the *Membranipora* colonies grow onto the new laminarian fronds, and survives until the following winter.

Pycnogonids are frequently found on bryozoan colonies, and are assumed to be predators. The relationship has, however, been studied only in the Antarctic species *Austrodecus glaciale*.[44] Pycnogonids which feed on actinians and hydroids have short, wide, blunt-ended proboscides. That of *A. glaciale*, on the contrary, is 2 mm long, extremely slender, down-curved and distally flexible; it can be moved in the ventral direction by a powerful muscle. *A. glaciale* appears to be a specialist predator on ascophorans, and is thought to gain access to the polypide via the pseudopores in the calcification (see Fig. 14A,B, p. 96). A styliform proboscis occurs in several species of *Austrodecus* and in some other genera; none of these species, nor those having short proboscides but nevertheless found on bryozoa, have had their feeding habits investigated.

GROWTH AND REPRODUCTION

Life cycles and breeding seasons

Breeding seasons can be determined in a majority of bryozoans simply by observation of the colonies. Developing ova and embryos are conspicuous and their abundance can be estimated. An alternative method, employed by biologists concerned with marine fouling, is to make use of experimental test surfaces on which month by month settlement can be recorded. It should be noted, however, that the presence of embryos in ovicells during the winter does not imply that larvae are being released.[23,42]

The life span in bryozoans in cold-temperate waters varies considerably, for colonies may be short-lived, annual, biennial or perennial. Many epiphytic species, such as *Callopora lineata* and *Hippothoa hyalina*, have a brief life cycle with several overlapping generations each year.[42] In effect there is a prolonged breeding season during which a growing algal frond may be colonized. In short-lived species the proportion of zooids bearing embryos is high and larvae are produced in quick succession by any zooid, roughly fortnightly in *Callopora dumerilii*.[88]

In *Bicellariella* there are two generations each year.[42] Ancestrulae and tiny colonies produced from an autumn settlement overwinter, grow rapidly in spring, produce larvae by June and then die. The offspring of this generation grow and mature very quickly (it has been reported in *Bugula neritina* that under warm summer conditions colonies achieve maturity at a smaller size than in the spring[23]) and die before the onset of winter, having produced in early autumn the larvae which give rise to the over-wintering stock. *Bugula flabellata* has likewise two generations each year;[42] the colonies which disintegrate as winter approaches may not die, however, but survive as dormant stolons and give rise to fresh growths in the spring.

An annual cycle associated with a clearly defined breeding season occurs in *Crisia eburnea*. The old stock dies before the end of the year and young but well-grown colonies survive the winter. More commonly (*Escharina spinifera*, *Fenestrulina malusii*), a proportion of the old colonies survives the winter, so that the species is partly annual and partly biennial. This may result in a longer breeding season with old or well-grown colonies maturing well before those which overwintered as ancestrulae or little more.

Other bryozoans, of which *Flustra foliacea* is a familiar example, are perennial. Somatic growth takes place from spring to autumn: during the latter season eggs and sperm are produced. Transfer of eggs to the ovicells commences in October, and here the embryos remain until their release as larvae in early spring. Colonies may survive for at least ten years, adding a 2–4 cm zone to their fronds each year. One splendid colony weighed 13 g and was estimated to consist of $1{\cdot}3 \times 10^6$ zooids;[16] and it must also have been a large colony which, when placed in water by Sir John Dalyell,[14] liberated 'at least ten thousand' larvae. Less protracted brooding of embryos is more usual, and in other perennial species larvae are liberated during the summer. In most of these (e.g. *Escharoides*, *Porella*, *Schizoporella*) only a few zooids at a time produce embryos. Most perennial species display little clear seasonal variation in growth and reproductive rates, although there may be a slow growth period during the winter.

The species with a clearly defined reproductive season appear to be on the whole those with a restricted boreal or temperate distribution. The former group, being in Britain towards the southern limit of their distribution, breed in winter–spring; the latter group, here reaching their northern limit, breed during the summer–autumn. Widely distributed species tend to have a less marked reproductive season.[42,45] Overall, however, in temperate latitudes, reproductive activity is high during summer and autumn with the peak falling in August–September. Since the pattern of activity so clearly follows the annual variation in sea temperature (Fig.10), and breeding season lengthens with decreasing latitude (Fig.11), it is tempting to suggest a causal relationship. Such dependence may well exist, but it is not the only factor involved, because a similar annual cycle is shown by bryozoans living deep in Norwegian fjords where temperature fluctuates by less than 2°C.[23] Moreover, in temperate latitudes many species resume growth or reproductive activity in early spring before the ambient sea temperature has started to rise.[42,47]

Observations of this kind suggest that a correlation with day-length, with planktonic production cycles or with less obvious envi-

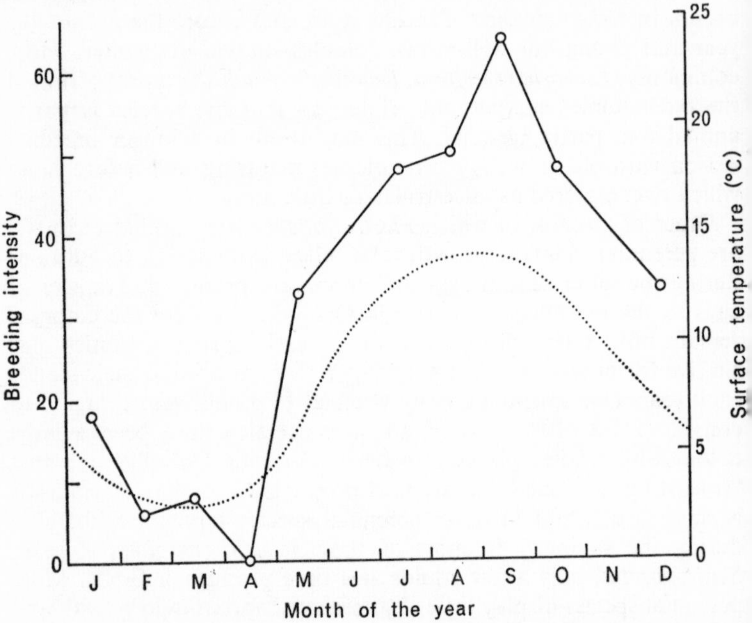

Fig. 10 Seasonal reproductive activity

The graph shows the seasonal pattern of reproductive activity in bryozoa from the west coast of Norway (latitude 60°N). Left ordinate is arbitrary index of the abundance of developing embryos. Dotted line shows sea surface temperature.

ronmental changes should be sought. Unfortunately the lack of data on the food of bryozoans is matched by the dearth of information about production peaks of particular species among the smaller phytoplankton. Important advances have, however, been made recently in compiling overall pictures of phytoplankton production.[153] In the most general terms it appears that, while in the oceanic, eastern North Atlantic there is a single phytoplankton peak in the late spring (end of April or May), in coastal waters there are two peak periods, March–April and September–October. The second outburst, which may be as great as or greater than the first, is to some extent the product of species whose 'flowering' comes late in the year, but is also, presumably, boosted by some of the same species which contributed to the spring outburst. Thus, near the European coast, phytoplankton is most abundant during spring and autumn.

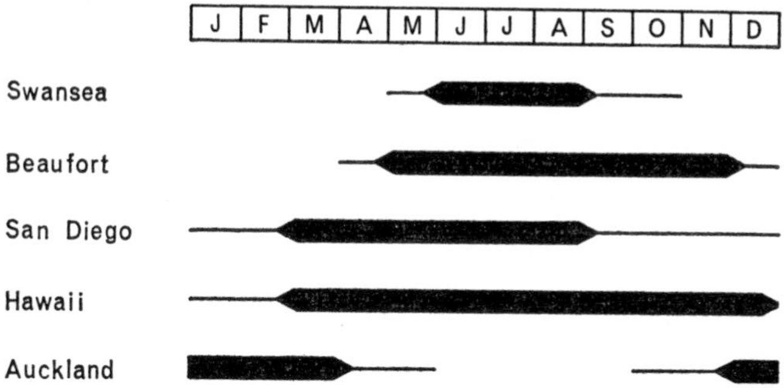

Fig. 11 Settlement of *Bugula neritina* in different parts of the world

The diagram shows how latitude influences the duration of the breeding season in a cosmopolitan species. The localities are:

Swansea (South Wales, U.K.) 52°N
Beaufort (North Carolina, U.S.A.) 35°N
San Diego (California, U.S.A.) 33°N
Hawaii 21°N
Auckland (New Zealand) 37°S

The available evidence suggests that the commonest form of bryozoan annual cycle in temperate latitudes is characterized by rapid somatic growth in the early part of the year, although the whole subject of colonial growth requires thorough investigation (cf. p. 154). Reproductive activity is maximal late in summer and declines steadily as autumn progresses. Temperature and a day-length factor—presumably the production of phytoplanktonic food—govern the processes, but the roles of each factor cannot yet be analysed, chiefly because the detailed patterns found in bryozoans vary so much from species to species.

5

MORE ABOUT THE CHEILOSTOMATA

The cheilostomes dominate the Gymnolaemata and constitute the most successful group of living bryozoa. The order was introduced in Chapter 2, but consideration of certain characteristics was then deferred. Among these are the striking polymorphism of zooids; a system providing communication between zooids which, although imperfectly understood, clearly plays an important role in determining the way in which colonies develop, and without which the evolution of specialized non-feeding zooid morphs would have been impossible; a special type of brood chamber; and a remarkable way of thickening the zooid wall. These four features, all of which appear to have been influential in promoting the group's success, form the subject matter of this chapter.

INTERZOOIDAL CONNEXIONS

Cheilostome zooids, like those of ctenostomes, are connected with each other by funicular material passing through mural pores of various kinds.[89] The simplest condition, seen for example in *Labiostomella*, resembles that found in ctenostomes like *Paludicella*. A new zooid arises as a distal swelling which becomes separated from its parent zooid by a transverse septum (Fig.12A). The latter is at first annular, but extends centripetally until there remains only the small central pore through which the funiculus passes. In many cheilostomes, however, the transverse septa calcify around a number of small pores rather than a single large one.

A more characteristic cheilostome zooid may be likened in plan view to a hexagon. The three proximal facets of a zooid forming at the colony margin are already in contact and communication with existing zooids; the three distal facets are free to produce buds. The situation is most clearly understood in *Beania* (Fig.12B) because its zooids are not, as in most cheilostomes, close together, but are spaced out and joined to one another by short interconnecting 'tubes' (extensions of the zooids). Normally the bud arising from the terminal facet forms a tube which meets and merges with its counterparts from the nearer disto-lateral facet of each of the zooids on either side of it (zooids 4, 5 and 6 in Fig.12B). Thus three buds fuse to form a daughter zooid (7 in Fig.12B), and each completed zooid in turn gives rise to three more buds. Every bud becomes cut off from its parent zooid by a porous septum of the type just described. When a new line of zooids is initiated, the terminal bud of the parent zooid continues the existing line, while one of the disto-lateral buds becomes the proximal tube of the first zooid in the new line.

In the great majority of cheilostomes the lines of zooids are pressed against each other so that a continuous crust results. The zooids may still be viewed as hexagons or rectangles which are alternately or quincuncially arranged, as bricks in a wall (Fig.12D). New zooids arise terminally and are separated by perforated partitions exactly comparable to the septa already mentioned. Just as in *Beania*, a new line of zooids can arise from a disto-lateral bud (Fig.12D, zooids 1 and 3); but this happens infrequently in the zone of astogenetic repetition, except in rapidly expanding circular colonies, since adjacent lines of zooids already occupy all the available space.

It follows from the manner in which the colony is made up that the lateral dividing walls between component lines are double structures formed equally by the zooids of the two contiguous lines (Figs.12C,14B,C). This duplex condition contrasts with the unified structure of the transverse septa, and explains why colonies will often separate into lines of zooecia when treated with hypochlorite solution.[28]

The presence of adjacent zooids does not prevent the formation of buds on the disto-lateral facets, but does dictate the manner in which they develop. At the centre of the facet a porous septum, concave with respect to the parent zooid, is formed, exactly comparable to the septa already mentioned (Fig.12D,E). This porous septum is variously called a *pore-plate*, *rosette-plate* or *septulum* (dim. of *septum*). Outside the pore-plate, the cuticles of the two wall components separating the contiguous zooids break down to produce a large single opening (Fig.12F¹–F⁴). In this way communication between lateral zooids is established by buds whose growth into autozooids is prevented: each zooid is 'broken into' on its proximo-lateral facets, while itself 'breaking out' from its disto-lateral facets (Figs. 12D,14B,C).

A variation of this structure is the *pore-chamber* or *dietella* (dim. of *diaeta*, room) found in many cheilostomes. Here the porous septum, instead of joining the lateral wall all around the future mural pore, unites with the basal wall below the pore (Fig.12G). Pore-plates and pore-chambers are thus tiny zooid buds which, for lack of space, have become interzooidal connexions instead of autozooids (see Fig.3B,C, p. 33).

Pore-chambers can also occur in association with transverse walls, as in *Callopora*, but develop in a rather different manner from that just described. The true septum is the porous wall of the chamber; but a second partition, an invagination of the exterior walls, arises distally to it (Fig.12H), and closes centripetally to leave only a large single opening (see also Fig.3A and the right-hand zooid in Fig.5A, p. 39). The pore-chamber is now complete. The next daughter zooid in the line swells as a bud out of this large pore (Fig.12H, extreme right).

Erect unilaminar colonies—such as those of cellularioids *Bugula* and *Notoplites*—can also develop pore-plates in the basal zooidal walls. From the pore-plates tubular kenozooids may arise. Some of

---

Fig. 12 Interzooidal connexions in gymnolaemates

The diagrams show how the gymnolaemate colony is built up from lines of interconnected zooids, and how lateral communication between such lines is achieved. A full explanation is given in the text

A *Paludicella*: zooids in single series
B *Beania*: zooids quincuncially arranged and connected by 'tubes'
C Representation of a cheilostome colony with an ancestrula (*A*) which has given rise to three buds and, initially, three series of zooids. Communication pores have not been shown in the duplex walls
D Disposition of pores and pore-plates in quincuncially arranged zooids, and the inception (zooids 1, 2 and 3) of a new series of zooids. (Each facet commonly bears two pore-plates although only one is shown)
E A single pore and pore-plate from D in vertical section. The pore-plate (white) is a septum in the distal half of its zooid, the wall of which (black) contributes to the pore together with a wall in the proximal half of the contiguous lateral zooid (hatched)
F Four stages in the development of a septum (pore-plate) and pore of the kind shown in E. (A somewhat different method of formation has been reported by Banta[28A])
G Lateral pore-chamber (*c*). Conventions as in E
H Transverse (morphologically proximal) pore-chambers (*c, c'*) in a line of zooids; a zooid bud is developing from *c'*.

(Based on Silén[89])

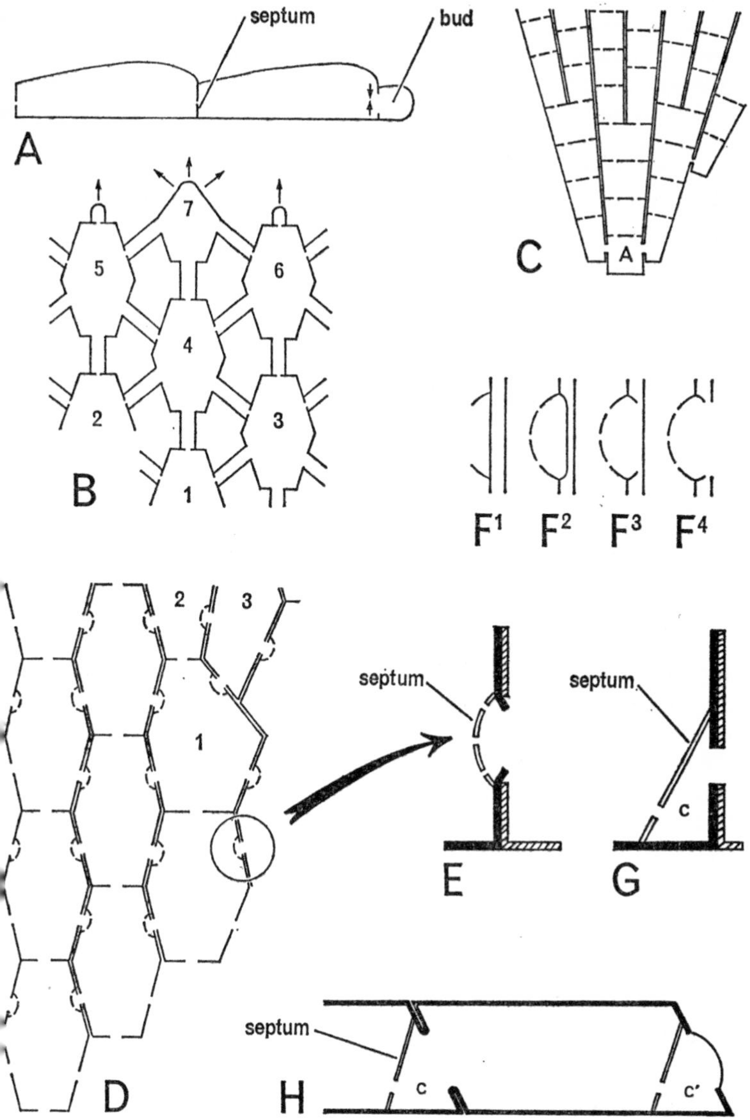
septum
bud
A
B
C
1
2
3
4
5
6
7
A
F1
F2
F3
F4
D
E
G
H
septum
septum
c
septum
c
c'

Fig. 12

these, which display negative phototropism in *Bugula*,[80] grow downwards to become the rhizoids that anchor the colony to the substratum. Others, especially at branch bifurcations in *Notoplites* for example, connect the pore-plate of one zooid to that of another, usually according to some clearly apparent pattern. Occasionally, as Dr Hastings shows in several illustrations in her 'Discovery' Report,[54] a rhizoid from one branch may grow towards and join with the pore-plate of a zooid in another branch. Such oriented growth towards a pore-plate might be achieved by chemotropism, but evidence for the mechanism is at present entirely lacking.

POLYMORPHISM

One of the most striking characteristics of bryozoa is the extensive occurrence of polymorphism among zooids. One variant, the kenozooid, has already been mentioned. It occurs in both of the gymnolaemate orders and in the Cyclostomata, and is essentially a zooid morph devoid of internal structure, which contributes to stolons, to the attachment rhizoids of cellularioids, and which forms spines and various other empty chambers. Two more imposing kinds of heterozooid are the *avicularium* (*avicula*, dim. of *avis*, bird) and the *vibraculum* (*vibrare*, to move to and fro), both of which are restricted to the Cheilostomata. Avicularia are generally smaller than autozooids and lack a functional polypide. Their most characteristic feature is the *mandible*, a movable portion of the zooid which can be opened or closed like a jaw by the action of well-developed muscles. In vibracula the mandible, with its movements restricted to one plane, is replaced by a more freely articulated *seta*. A comparative study of their various types soon reveals why avicularia and vibracula are restricted to the Cheilostomata: the mandible and seta respectively are the homologues of the operculum of the autozooid.

Fig. 13 Cheilostome heterozooids

A Transverse optical section through a vibraculum
B Sessile avicularium, mandible closed
C Sessile avicularium, mandible open
D Sessile avicularium, frontal view
E Pedunculate avicularium of *Bugula*

add adductor muscle; *abd* abductor muscle; *crypt* cryptocyst; *f m* frontal membrane; *ins* points of insertion of the adductor muscle tendons; *mand* mandible

(A–C based on Marcus[64, 65])

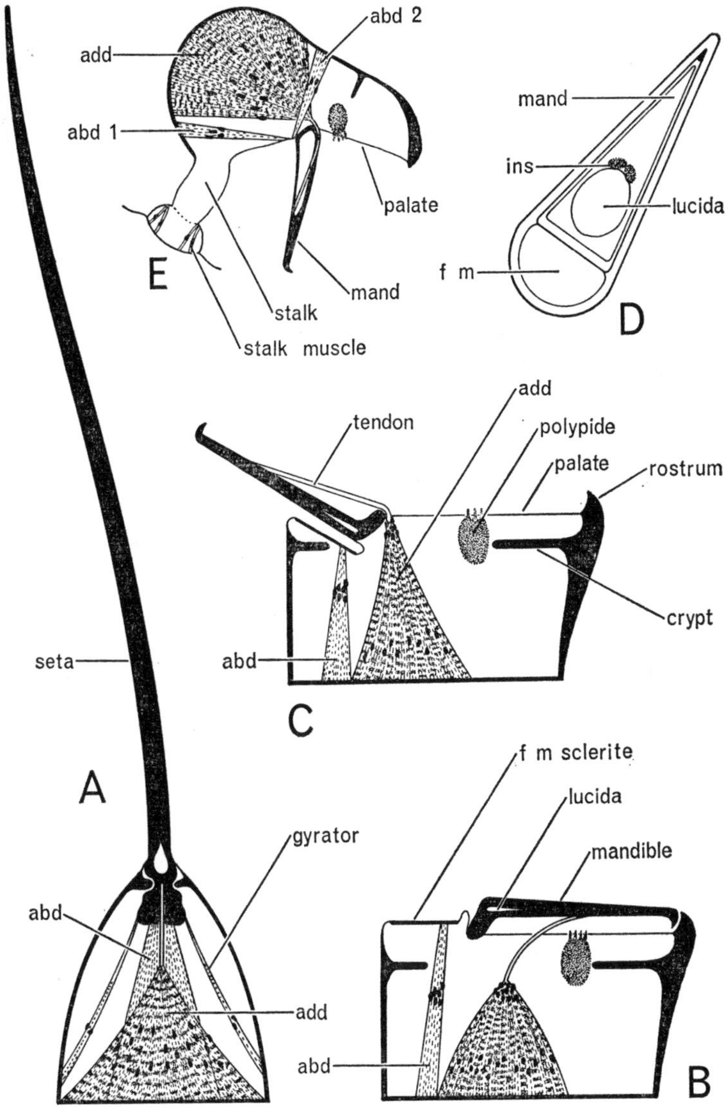

add
abd 2
abd 1
palate
E
mand
stalk
stalk muscle
mand
ins
lucida
f m
D
add
tendon
polypide
palate
rostrum
crypt
seta
abd
C
A
f m sclerite
lucida
mandible
gyrator
abd
add
abd
B

Fig. 13

Avicularia

The basic structure of the avicularium is that of a cystid with an enlarged and generally tapering operculum (mandible) hinged at its base to the frontal wall (Fig.13B–E). The mandible closes into a matching space, equivalent to the orifice, termed—to complete the analogy with jaws—the *palate*.[14] A polypide rudiment, thought to have a sensory function, may lie just inside the palate. The mandible is made up of two chitinous walls separated in the centre of the proximal half by a shallow space. This cavity shows in frontal view as a less opaque region of the mandible, and is termed the *lucida* (*lucidus*, lucent). A pair of projections, which may be united to form a bar, reinforces and delimits the palate at the line of the mandibular hinge. Proximal to this bar is an uncalcified area equivalent to the frontal membrane of the anascan autozooid. The coelomic cavity of the avicularium is largely filled by the mass of the paired occlusor muscles, now called the *adductors of the mandible*. Each muscle, from its wide origin on the basal wall, tapers to a tendon which inserts on a strengthened part of the mandible. Sometimes the two muscles converge to a single tendon. There may be one or two pairs of smaller *abductor muscles* which insert on the frontal membrane as do the divaricators of the operculum.

A study of heterozooids often begins with the stalked avicularium of *Bugula* (Fig.13E; also Fig.5E, p. 39), despite the fact that in the cellularioids we see avicularia in their most complex and highly evolved state. Here we shall consider a morphological series which culminates with the stalked avicularium, trying to gain some insight as to how these intriguing structures might have evolved.

A starting point is found in the genus *Steginoporella*, in which autozooids of two types are found, generally distinguished as A and B forms. Both have functional polypides, but the B zooid has a larger operculum and better developed muscles for operating it. The B zooids suggest the stage of incipient avicularia, although the condition described for *Steginoporella* is apparently the result of a trend through the Caenozoic of decreasing dimorphism. Greater dissimilarity is seen in *Crassimarginatella* with the avicularium, as it may now be called, distinguished by having a hypertrophied operculum, although there is still a functional polypide.

The autozooids in a cheilostome colony are often arranged quincuncially. It is this regular arrangement that makes it obvious that the avicularium of *Flustra*, for example, simply replaces an autozooid in the linear series of zooids. Such avicularia are called *vicarious* (*vicarius*, taking the place of something). They are very slightly smaller than the normal zooids, and are characterized by the larger operculum and the lack of a polypide. Species of *Cellaria* have vicarious avicularia which differ markedly in shape from the autozooids.

As growth proceeds in a circular or expanding colony, the number of radiating series of zooids must increase in proportion to the ever lengthening circumference. Thus, at intervals, a zooid gives rise to a daughter zooid in a disto-lateral position as well as terminally, thereby initiating a new series of zooids (Fig.12D, zooids 1–3). Vicarious avicularia frequently arise at such bifurcations.[84] In some of the Flustridae (*Chartella*, *Securiflustra*) it is the terminal bud which forms the avicularium, and the disto-lateral bud which becomes an autozooid; in others of the family (*Flustra*, *Sarsiflustra*) the relations are reversed, so that the avicularium forms disto-laterally. Similarly Schager[1] has observed that in the genus *Floridina* it is the disto-lateral bud which becomes the avicularium. The difference between the terminal and disto-lateral positions is that the main funiculus of a zooid passes through the pore-plate of the former.

The same situation occurs in *Thalamoporella*, where only the disto-lateral bud at a bifurcation may develop into an avicularium. Powell and Cook[70] have noticed, however, that if the disto-lateral daughter zooid grows as rapidly as the terminal daughter zooid, it likewise forms an autozooid. Apparently it produces an avicularium only if the space available for its development is restricted, and its potential for becoming an autozooid cannot be realized.

The next stage in our morphological series shows avicularia no longer obviously replacing autozooids, but being smaller and wedged in between them, as in *Escharina*. Such avicularia are termed *interzooidal*, and the example of *Thalamoporella* just discussed suggests one of the possible mechanisms by which they could have evolved.

In a majority of cheilostomes the avicularia, now much reduced in size, are attached to the frontal or lateral walls of the zooid: they are then termed *adventitious* (*adventicius*, coming from without). Possibly the evolutionary process has continued along the lines suggested above, so 'squeezing out' interzooidal avicularia until they are actually supported by another zooid; or they may represent a true frontal bud, since these are known to occur in the Cheilostomata.[84] Adventitious avicularia sometimes develop from marginal pseudopores (see p. 98).

In anascans, adventitious avicularia occur most frequently on the gymnocyst just proximal to the frontal membrane (Fig.5A,B). When brood chambers or *ooecia* are present, presumably because they often rest on the proximal part of the next zooid, the avicularia may be situated atop them. In *Cauloramphus* one or more of the spines which encircle the frontal membrane is replaced by an avicularium. In ascophorans adventitious avicularia may be attached almost anywhere, but two arrangements are particularly frequent. There may be one in the 'suboral' position just proximal to the orifice or there may be a pair of avicularia, one each side of the orifice (Fig. 5H,I). The avicularia found in a colony of any species may be of more than one kind.

The usual orientation of the adventitious avicularium is with its basal surface embedded in the frontal wall of the bearing zooid. It may be rather flat in shape, or its cystidial space may be inflated to form a *chamber*, raising the avicularium a little from the plane of the frontal wall. The mandible is usually either triangular or rounded.

In *Bugula* the proximal part of the avicularium is slender, forming a stalk, which not only raises the avicularium above the frontal membrane, but provides it with mobility, for the stalk is jointed and contains muscle fibres (Figs.5E,13E). The morphologically basal part of the avicularium is wholly free from the bearing zooid, and is drawn out above the palate into a beak or *rostrum*, below which is the mandible. Here are the birds' heads which gave rise to the word avicularium; for, as Darwin wrote when discussing them in *The Origin of Species*, 'they curiously resemble the head and beak of a vulture in miniature, seated on a neck and capable of movement'.

It is generally supposed that avicularia serve a protective function and are analogous to the pedicellariae found in echinoids. Small predators and larvae seeking a location for metamorphosis may be discouraged, and it is not uncommon to find small nematodes grasped by the better developed avicularia. Harmer[14] recorded that in *Bugula* small polychaetes and crustaceans may be captured, and Kaufmann[1] noted that the avicularia are discouraging to small tube-building amphipods (e.g. *Jassa*) which are found among the colonies. In physical terms the mechanical advantage of the adductor muscle increases *pari passu* with the degree of closure of the mandible, so that avicularia will be most effective against small or slender objects. The appendages of small crustaceans, and perhaps of pycnogonids, come into this category. It is unlikely to be coincidence that the most elaborate and mobile avicularium known has developed in a group, the Cellularioidea, which retains a large and generally otherwise unprotected frontal membrane. The family Bugulidae, which above all displays the stalked avicularium (although it is absent in *Bugula neritina*), is one of the most successful anascan groups.

Vibracula

Occasionally the mandible of an avicularium is very long and slender, as in *Microporella*, and this suggests the way in which vibracula might have arisen. *Crepidacantha*, in particular, has a heterozooid which appears to be intermediate between an avicularium and a vibraculum.[55] The mandible is setiform, but can move only in one plane like that of an avicularium. However, it is not hinged at its base but swings between a pair of supporting condyles just above its lower extremity. This arrangement approaches that found in vibracula.[64,65]

In gross terms a vibraculum consists of a basal chamber from which springs a long tapering seta (Fig.13A). The articulation appears to be a development of the kind described in *Crepidacantha*, for hollows in the base of the seta are aligned with a pair of asymmetrical condyles. The arrangement is such, however, that the seta can to some extent swivel about its own axis. The chamber contains equal sized adductor and abductor muscles, and smaller *gyrators*.

Vibracula occur in *Scrupocellaria* and certain other genera of the Cellularioidea and in some of the Coelostegoidea. In *Scrupocellaria*, species of which occur commonly all over the world, the sweeping movements of the setae can readily be observed. The function of vibracula is thought to be that of discouraging small organisms, especially larvae searching for somewhere to settle, on the surface of the bryozoon. They may also, as explained in Chapter 4, remove inanimate particles, especially in the non-attached colonies of *Cupuladria* and *Discoporella* which live on a sandy sea bed.

OOECIA

It seems likely that the original reproductive system in gymnolaemates involved the production of small eggs developing into cyphonautes larvae; but this was replaced during the evolution of many species by the telolecithal brooded ovum. Incubation in an embryo sac suspended in the coelom might have been the primitive arrangement but, since the embryo then necessarily fills much of the cystid, the polypide degenerates. This circumstance may have favoured the evolution of external brood chambers. Most often in cheilostomes the ovum develops in a special calcified brood chamber, the *ooecium* or ovicell (Figs.3A, p. 33; 5A,B,C,E,F,G,I, p. 39; 14A; 15E). In its commonest form the ooecium is a hood-like structure attached to the distal end of one autozooid, and resting on or embedded in the next. As ooecium formation, in some species at least, is coincident with the appearance of the first ovum in the ovary, its commencement may be a response to the secretion of some ovarian hormone.[88]

The ooecium arises from the position of the distal zooid wall as a pair of flat outgrowths joined in the midline.[58] Possibly these represent a pair of much modified spines.[50] They spread outwards and upwards, and then overarch to enclose an almost spherical cavity. The wall of this chamber is double,[88] with an outer calcified layer or *ectooecium* separated from an inner membrane or *entooecium* by a narrow coelomic space (Figs.14A;15E).

As the ooecium develops, its lumen becomes partly filled by an *inner vesicle*, a membranous evagination from the distal part of the zooid. The ovum, as it runs out from the supraneural pore, re-forms in the space between the entooecium and the inner vesicle: here it develops (Fig.14A).

The external opening of the ooecium is often situated just above but perpendicular to the orifice of the parent zooid (Fig. 3A). Such an ooecium is described as *hyperstomial*. Where the ovicell is at its simplest, as in *Callopora* (Fig.5A,B), its opening is closed only by the inner vesicle, the exposed part of which may be chitinized as in *Scrupocellaria*.[86] Presumably to provide better protection for the embryo, in such cheilostomes as *Smittina*, the ooecium wall may extend some way over the orifice of the parent zooid. The arc described by the free margin of the operculum as it swings between open and closed limits the extent of such development. Ryland[1] has observed that in *Pentapora*, which has this kind of ovicell, the operculum actually has two closed positions: one fitting into the orifice proper, thereby leaving the ooecium open; the second sealing the *common orifice* of both zooid and ooecium. Finally in many genera, both anascan and ascophoran, e.g. *Micropora*, *Membraniporella*, *Haplopoma*, the occluded operculum always seals the common orifice (Fig.14A). An ooecium closable in this way is described as *cleithral* (*kleithron*, a means of closure).

Cleithral ooecia are more or less embedded in the substance of the

---

Fig. 14 Zoid structure in the Ascophora and Gymnocystidea

A Optical section of zooid with tentacles retracted. (The wall type illustrated is a tremocyst, and the ooecium is cleithral)

B Transverse section through the proximal part of A

C Transverse section through the distal part of a gymnocystidean zooid.

Calcitic primary walls of adjacent zooids are shown respectively black and hatched; aragonitic secondary calcification is shown stippled.

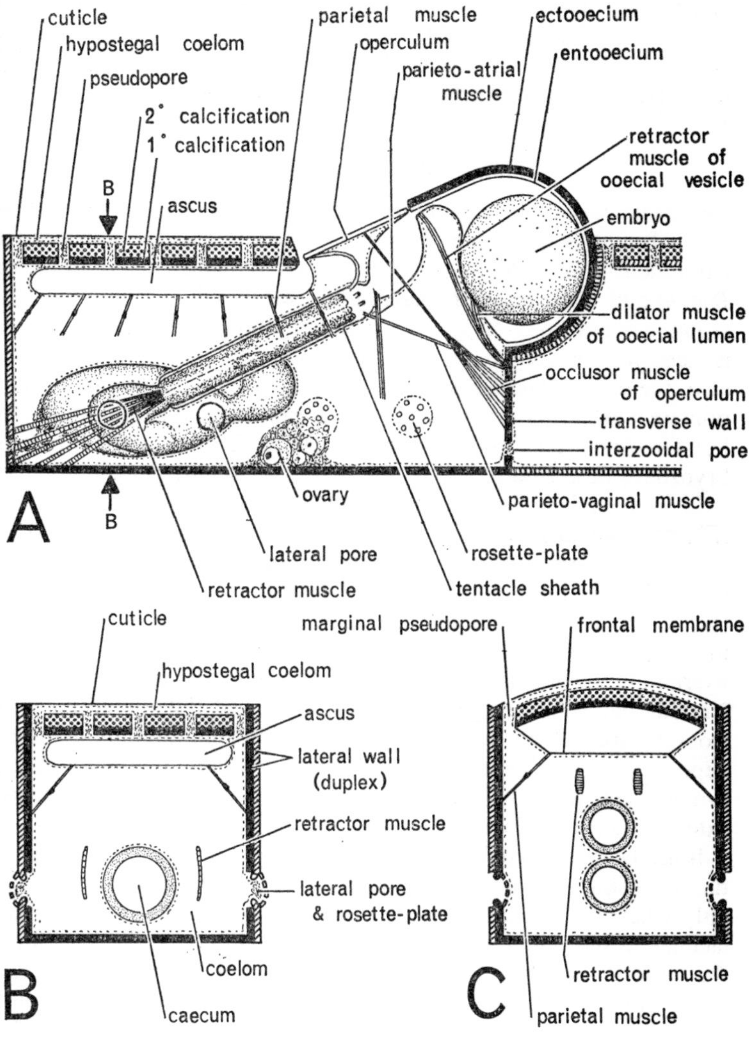
cuticle
hypostegal coelom
pseudopore
2° calcification
1° calcification
B
ascus
parietal muscle
operculum
parieto-atrial muscle
ectooecium
entooecium
retractor muscle of ooecial vesicle
embryo
dilator muscle of ooecial lumen
occlusor muscle of operculum
transverse wall
interzooidal pore
parieto-vaginal muscle
ovary
B
A
lateral pore
rosette-plate
retractor muscle
tentacle sheath
cuticle
hypostegal coelom
ascus
lateral wall (duplex)
retractor muscle
lateral pore & rosette-plate
coelom
caecum
B
marginal pseudopore
frontal membrane
retractor muscle
parietal muscle
C

Fig. 14

distal zooid. The culmination of this trend finds the ooecium wholly immersed in the distal zooid: it is then termed *entozooidal* (Fig.15E). Whatever the type of ovicell, it may be yet further protected by secondary calcification which spreads over the ectooecium.[69]

SECONDARY CALCIFICATION

The zooid walls of ctenostomes are uncalcified, though not necessarily devoid of calcium carbonate. In cheilostomes the walls, apart from the frontal membrane where present, are usually calcareous. In *Bugula* and other anascans having the same colony form, the dry skeleton contains 25–50% organic matter, and the calcium carbonate is deposited as calcite. In other cheilostomes, and in cyclostomes also, the organic matter comprises less than 25% by weight.[82] Most non-erect anascans deposit skeletal material in the form of calcite,[70A] but Greeley[106] reports that colonies of the free-living lunulitiform bryozoans described in Chapter 4 also contain aragonite.

In some of the less complex ascophorans such as *Hippothoa* (Fig.5G), the primary wall which forms the front of the zooid is smooth and imperforate; it does not become covered by secondary thickening during the life of the colony and is termed a *holocyst* (*holos*, entire). More commonly, however, the primary frontal wall in ascophorans and the frontal shield in gymnocystideans becomes overlain by secondary calcification. Rucker[1,70A] has recently demonstrated that whereas the primary walls—basal, lateral, distal and frontal—are usually calcitic, secondary thickening is more often deposited in the form of aragonite. How, it will be wondered, can a primary calcitic wall secreted by an underlying epidermis become externally covered by aragonite?

It has been realized for some time, notably by Canu and Bassler,[100] that deposition of secondary calcification in ascophorans takes place only when the primary skeletal layer contains perforations. These lacunae in the calcification, or *pseudopores*, do not in fact penetrate the cuticle and are plugged with tissues from the underlying cell layers (Figs.5F,H,I;14A–C). As envisaged by Canu and Bassler, but only recently shown even to be possible, the cells plugging the pseudopores delaminate above the primary skeletal layer to form a second epithelium. It is this which deposits, from its inner surface, aragonite in contact with the calcite layer. The aragonitic and calcitic laminae do not bond together, and sometimes separate in fossilized material.

Outside the secretory epithelium is a narrow coelomic space or *hypostegal coelom* (*hypo*, under+*stege*, roof), lined with peritoneum and bounded externally by an epidermis and cuticle.[83] It will be explained in Chapter 6 how some of the groups of Stenolaemata have also evolved a complex wall of this type, incorporating a hypostegal coelom, but formed by an entirely different method in which pseudopores are not involved. The structure described in Chapter 2 for coelostegoideans, in which a cryptocyst underlies the frontal membrane, is also similar in that a hypostegal coelom is present.

The pseudopores provide the key to understanding wall structure in the Ascophora, for they facilitate coelomic continuity. The aragonitic layer is deposited around the tissue columns penetrating

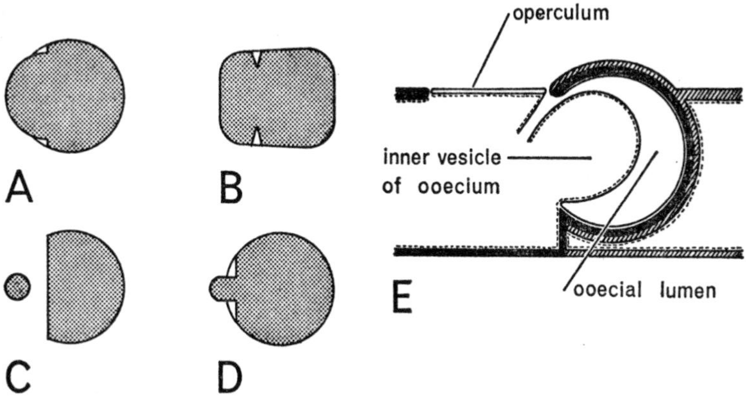

Fig. 15 The orifice in ascophorans, and the entozooidal ovicell

A–D Types of orifice found in the Ascophora (A, B and D are selected examples, between which many intermediates occur):
- A Bipartite orifice with large anter and small poster
- B Orifice with anter and poster separated by pointed condyles
- C Microporellid orifice with separate ascopore
- D Schizoporellid orifice with narrow poster (sinus)
- E Section through an entozooidal ooecium (compare Figs 3A and 14A)

the pseudopores, so that the pattern of perforations persists however thick the wall becomes.

Two fundamental types of secondarily thickened frontal wall have been recognized in ascophorans, based on the distributional pattern of the pseudopores. Where these completely cover the front in a more or less regular manner, the wall is termed a *tremocyst* (*trema*, hole) and the secondary layer is deposited evenly over the zooid. Some genera having tremocystic walls are *Cryptosula*, *Schizomavella*, *Schizoporella* (Fig.5I), *Smittina* and *Watersipora*. In other ascophorans the pseudopores are confined to the periphery of the wall, so that they are known as *marginal pores* or *areolar pores*: the wall itself is then a *pleurocyst* (*pleuro*, side), and the secondary layer quite obviously spreads inwards from the line of marginal pseudopores. Examples are *Escharella*, *Pentapora*, *Porella* (Fig.5H) and *Smittoidea*.

The binary classification of secondarily thickened frontal walls may well be an over-simplification. The pseudopores in *Watersipora*, for example, are remarkably large compared with most others, and it is possible that there is really more than one kind of tremocyst; and is the pleurocyst in gymnocystideans like *Escharoides* (Fig. 5F; 14C) and *Umbonula* identical in form with that of an ascophoran like *Porella*? Fortunately the whole subject of ascophoran frontal walls is now receiving the attention from research workers that its complexity merits, and new discoveries are sure to be forthcoming.

6

STENOLAEMATA—PALAEOZOIC TO PRESENT

In this chapter the attributes of four bryozoan orders are discussed, the essential characteristics of which (Chapter 1, p. 26) link them within the class Stenolaemata. Three of the orders—Trepostomata, Cystoporata and Cryptostomata—are extinct, so that the extant Cyclostomata play a key role in facilitating our interpretation of the fossil record. The cyclostomes are, accordingly, described first. The methods employed by palaeontologists to elucidate the nature of fossilized remains are then briefly explained before proceeding to the accounts of the three extinct orders.

CYCLOSTOMATA

The cyclostomatous Bryozoa are characterized above all by the form of their autozooids. These are long and tubular, with a circular, terminal orifice (the name being derived from *kyklos*, round+*stoma*, mouth) and calcified, giving the colonies a distinctive appearance which readily identifies members of the order.

Structure and function in the autozooid

The body wall generally comprises peritoneum, epidermis and cuticle incorporating a calcified layer. The distal end of the zooid is closed, when the polypide is withdrawn, by the *terminal membrane*, which is a direct continuation of the zooid wall minus its calcified layer (Fig.16A). In the genera *Hornera* and *Lichenopora*, which have a more complex wall structure, the terminal membrane corresponds with the outermost layers only (Fig.16C). The structure of the terminal membrane is, therefore, exactly that of the anascan frontal membrane. Centrally the terminal membrane infolds as the *atrium*, leaving a tiny pore at the surface. In *Crisia* and related genera this pore is closed by a sphincter muscle. In all cyclostomes the lower end of the atrium can be constricted by the *atrial sphincter*, in the manner described earlier for *Bowerbankia*.

A structure not found in the zooids of any other living order is the *membranous sac*, comprising an epithelial layer on the inside of a basement membrane. This sac, which arises from the atrium at the distal edge of the sphincter muscle, totally encloses the polypide. Just proximally to the sphincter, the membranous sac is fastened to the zooid wall by a ring of eight short ligaments; thereafter it hangs freely in the coelom, rather close to the wall, except at the points of origin of the two retractor muscles and the funiculus (Fig.16A). There are no lateral funicular strands passing to the interzooidal pores as there are in cheilostomes.

The polypide is of the usual bryozoan type. The paired retractor muscles are shorter than in gymnolaemates; and the only other major muscles are the *atrial dilators*, which arise on the membranous sac just below the fixing ligaments, pass upwards between them, and divide into many branches which insert on the terminal membrane and atrium. It is interesting to note that if, as Harmer[50] suggested, the frontal membrane of cheilostomes is equivalent to the terminal membrane of cyclostomes—the zooids having become much wider and shorter—then the position and relationship with other structures of the parietal muscles and the atrial dilators exactly correspond.

The coelom comprises the usual two parts, minute mesocoel and much larger metacoel; but the latter is here divided by the membranous sac into two entirely separate units, an inner *entosaccal cavity* and an outer *exosaccal cavity*.[132A]

In the retracted zooid the dilator muscles of the atrium are relaxed and the sphincter is closed. The membranous sac is fully distended and, proximally to the fixing ligaments, lies close to the zooid wall (Fig.16A). When the atrial dilators contract and the sphincter relaxes, the atrium is drawn close to the body wall. The

Fig. 16 Zooid structure in the Cyclostomata

A Tubuliporoid zooid with tentacles retracted
B An adjacent zooid (middle region omitted) with tentacles expanded
C Coelocystic wall structure and orifice of *Hornera* (Cancelloidea)
ep 1 eustegal epithelium
ep 2 hypostegal epithelium
ep 3 zooidal epithelium

(A and C partly after Borg[119])

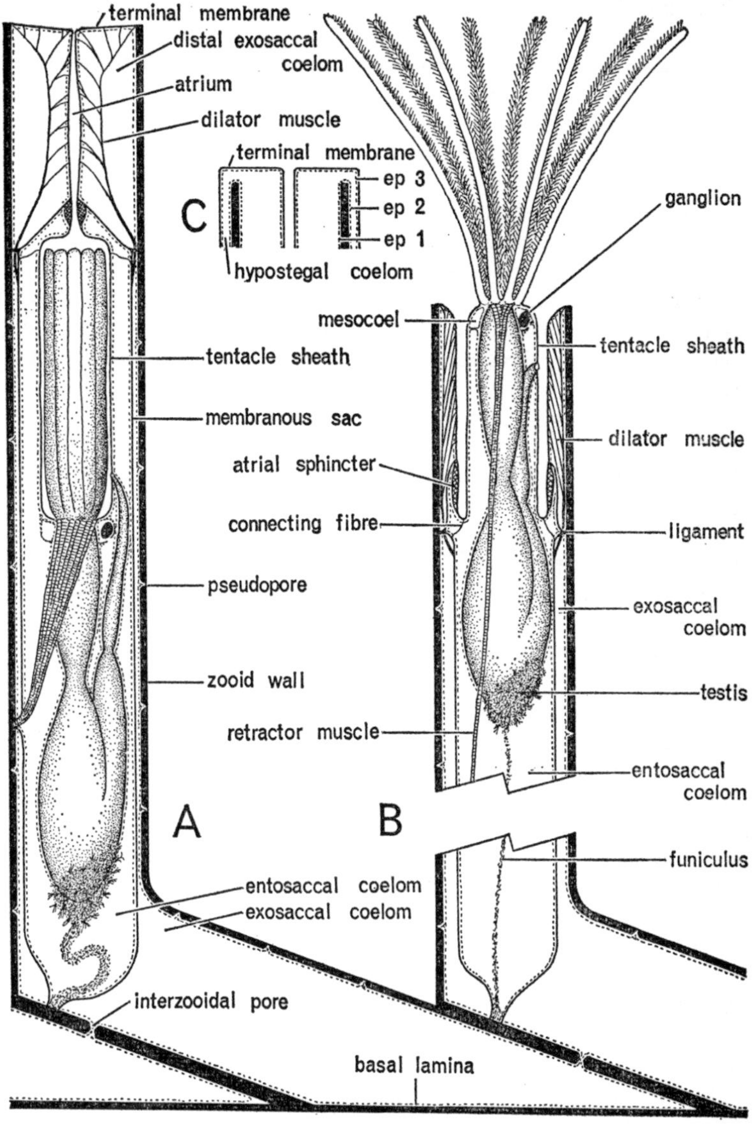
terminal membrane
distal exosaccal coelom
atrium
dilator muscle
terminal membrane
ep 3
ep 2
ep 1
C
hypostegal coelom
ganglion
mesocoel
tentacle sheath
tentacle sheath
membranous sac
dilator muscle
atrial sphincter
connecting fibre
ligament
pseudopore
exosaccal coelom
zooid wall
testis
retractor muscle
entosaccal coelom
A
B
funiculus
entosaccal coelom
exosaccal coelom
interzooidal pore
basal lamina

Fig. 16

fluid until then occupying the coelomic lumen surrounding the atrium is thus forced proximally, so compressing the membranous sac. As hydrostatic pressure in the sac rises the tentacle sheath starts to evaginate and in due course, as the retractor muscles relax, the tentacles emerge (Fig.16B).

In withdrawal of the tentacles, which is as usual a rapid process, the retractor muscles contract and the atrial dilators relax. Finally, when the tentacle sheath completely encloses the lophophore, the atrial sphincter tightens. The principles underlying the hydrostatic function of coelomic fluid are exactly the same in cyclostomes as in gymnolaemates, but the component parts through which the system is actuated are somewhat different. The true role of the membranous sac is obscure, for the system would appear to work just as well if it were absent!

Body wall

In a majority of Recent cyclostomes (e.g. *Crisia*, *Tubulipora*) the outer walls comprise the basic layers found in other calcareous bryozoa:[119] peritoneum, epidermis, calcified layer and hyaline cuticle, with numerous pseudopores penetrating the calcareous layer. The pseudopores create a punctate appearance which is most characteristic, so that the wall may be described technically as a *stictocyst* (from *stiktos*, punctured or spotted). In *Crisia* (Fig.17A), which is characterized by jointed, branching colonies, the calcified layer is represented in the joints by its non-limy matrix only.

Recent work on *Crisia* and other cyclostomes by Söderqvist[1] has added to our knowledge of the crystal structure of the calcified layer. Just inside the hyaline cuticle is a thin zone with finely prismatic microstructure, below which is a thicker laminated layer comprising superposed flat polygonal crystals. This layer deflects into the open ends of the pseudopores, indicating that the secreting epithelium continues into the pores as a living lining.

Transzooidal partitions or *diaphragms* may be found in cyclostomes, though never so abundantly as in the extinct order Trepostomata. Basal, intermediate and terminal diaphragms have been distinguished by Nye[1] according to their position within the zooid and the direction in which the calcareous laminae of the diaphragm flex when joining the inner zone of the lateral walls. The basal diaphragm is proximally situated with the laminae flexing orally, showing that the depositing epithelium must have been distal to the diaphragm (compare Fig.18B,D). The intermediate and terminal diaphragms, on the other hand, have proximally flexing laminae, indicating deposition by tissues aboral to the diaphragm.

The internal walls are single and shared by the contiguous zooids (and may therefore be termed *septa*), as in the Phylactolaemata but contrary to the condition found in most cheilostomes in which the lateral walls are of a duplex nature (Figs.12,14). Tubular pores, which widen towards each end, perforate the interzooidal walls and provide coelomic continuity (Fig.16A,B). These pores do not contain tissue plugs as in the Cheilostomata. True mural communication pores are unknown in the fossil remains of trepostomes, so that all living tissues must have been restricted to the oral side of the distal-most diaphragm. The interzooidal pore-system of the Cyclostomata maintains the inner part of the colony in direct connexion with the feeding zooids or their outermost, polypide-containing chamber. It would therefore be possible, Nye[1] points out, for the inner chambers to function as an extensive nutrient store providing against a period of adverse conditions, insuring the eventual reactivation of the colony. The older chambers of trepostome zooids could never have functioned in this way.

In three supposedly cyclostomatous families having extant members, exemplified by the genera *Heteropora*, *Hornera* and *Lichenopora*, the outside walls—with the exception of that which forms the base of the colony—are more complex than those of *Crisia* or *Tubulipora*. The living layers which line the zooid wall here continue on its outside as a double investment separated by a narrow extension of the exosaccal coelom; thus we have: peritoneum, epidermis, calcified layer, epidermis, peritoneum, hypostegal coelom, peritoneum, epidermis and cuticle (Fig.16C).[119] This kind of wall, which incorporates a coelomic space, will be called a *coelocyst*. The epithelium which lines the calcification internally is distinguished as *zooidal*, the remaining cell layers of the coelocyst are extrazooidal or *colonial*: the inner colonial epithelium will be termed *hypostegal* and the outermost, which underlies the cuticle, *eustegal*.

Heterozooids

Not all zooids in a cyclostome colony are autozooids. In *Crisia* there are anchoring *rhizoids* (*rhiza*, root); while in *Diplosolen* there exist miniature zooids of unknown function, each one complete with a rudimentary polypide possessing one tentacle. These are called *nanozooids* (*nanos*, dwarf). Above all, in many genera there are zooids modified for reproduction by loss of their polypides. Such *gonozooids* (*gonos*, offspring) act as chambers sheltering the developing embryos.

Reproduction

The homology of gonozooids is particularly obvious in the erect colonies of *Crisia*, where the pyriform brood chamber is clearly one of the series of regularly arranged zooids which make up the branches (Fig.17A). Relatively simple gonozooids are also found in *Tubulipora*, although they tend to develop into large and irregular structures fitted in among the autozooids. In *Lichenopora* the brood chamber is also a spacious structure, but only its proximal region originates as a fertile zooid; its main lumen is an extrazooidal space, the derivation of which will be explained later.

Germinal cells arise at the growing edge of the colony, and oogenesis apparently takes place there also. Most of the ova, however, degenerate; although a minority will be enveloped by the mesoderm cells of developing polypides. The mesoderm around the ovum constitutes a follicle, but no proper ovary is formed. As very few zooids are destined to become brood chambers, a majority even of those ova which have become associated with polypides will eventually abort. Those which survive become part of a gonozooid, the polypide of which degenerates, usually without having become functional, while the cystid enlarges to form the embryo shelter.

The male cells (spermatogonia) form a testis, which—uniquely among bryozoa—is enclosed within a cellular membrane, at the distal end of the autozooidal funiculus (Fig.16A). Another feature peculiar to Cyclostomata is that sperm develop in tetrads and not around a cytophore. When the nuclei of the spermatocytes divide, the cytoplasm does not. The four spermatids remain with their heads embedded in the common plasmic mass until the achievement of morphological maturity.[13] The formation of a cytophore appears to represent an extension of this arrangement.[119] The mature sperm are more primitive than those of other bryozoa in that the middle region is much shorter (see p. 49).

Reproduction in cyclostomes is chiefly remarkable for the occurrence of embryonic fission: a process which, as Harmer noted,

---

Fig. 17 *Crisia*, and cyclostome astogeny

A Part of *Crisia* colony
B Young primary zooid of *Crisia*
C–D Early astogeny of *Crisia*
E Early astogeny of *Tubulipora*, in surface view, with zooids numbered in order of formation
F Proancestrula of *Tubulipora* and *Lichenopora* (sectional view)
G Young primary zooid of *Tubulipora* and *Lichenopora*
H–I Early astogeny of *Tubulipora*
J–K Early astogeny of *Lichenopora*

al alveolus; *gz* gonozooid; *j* joint; *p* proancestrula; *rh* rhizoid

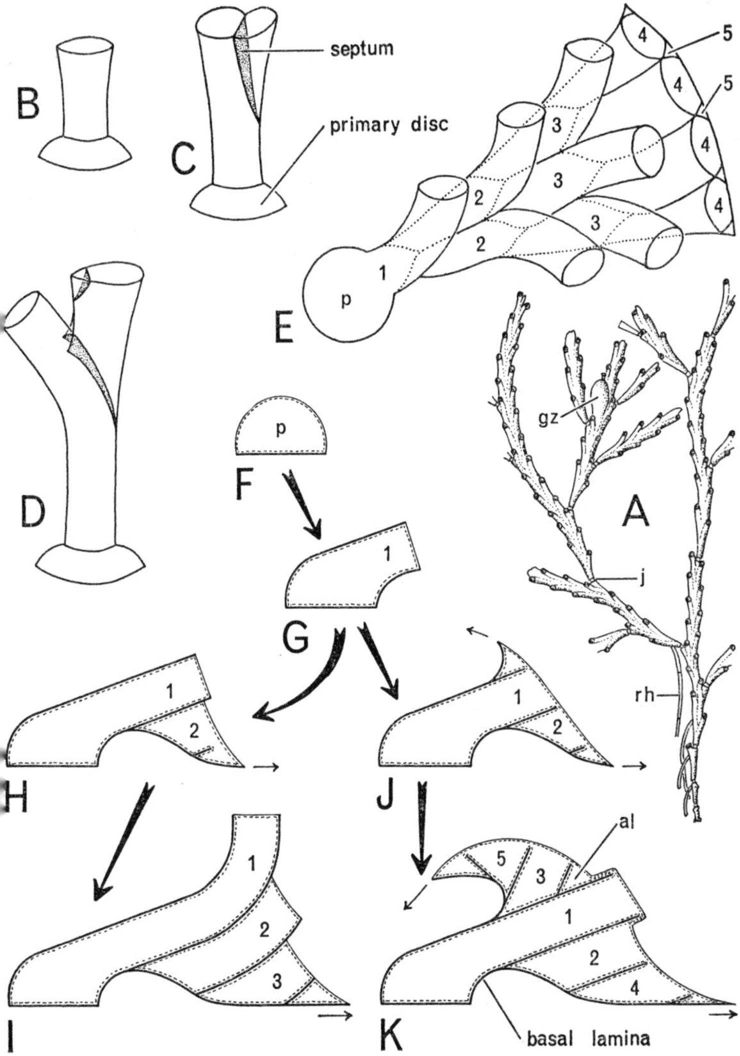
B
C
septum
primary disc
D
E
p
1
2
3
4
5
gz
A
j
rh
F
G
H
J
I
K
al
basal lamina

Fig. 17

appears to be without an exact parallel in the animal kingdom, although polyembryony in various guises occurs, for example, in trematodes, parasitic Hymenoptera and armadillos (Edentata). After fertilization has occurred, presumably in the manner now established for gymnolaemates (p. 50), although it has never been observed in the Cyclostomata, the zygote cleaves to produce an indistinctly hollow ball of blastomeres. This *primary embryo* then gives rise to lobate cellular processes which constrict off to form *secondary embryos*. These may in turn produce *tertiary embryos* in the same way, until a hundred or more have been produced. At this stage the brood chamber contains many solid balls of blastomeres, but the cells of each become rearranged in two layers around a central cavity. The embryos develop into ovoid ciliated larvae of simple structure, having only an apical invagination and an abapical adhesive sac (cf. gymnolaemate larvae, p. 51).[132A]

Astogeny in typical cyclostomes

The larva metamorphoses into a hemispherical primary disc or *proancestrula* (Fig.17F), which subsequently develops a tubular portion, thereby completing the primary zooid (Fig.17B,G). The polypide rudiment forms in the disc but extends upwards into the tubular portion as it develops. The tubular portion, whether erect or repent, has a morphological underside, the *basal lamina*. The distal edge of the basal lamina, where it merges with the terminal membrane of the zooid, constitutes the growing margin which gives rise to the entire colony.

The usual cell layers are recognizable at the growing edge of a cyclostome colony,[119] with the ectodermal cells cubical or cylindrical. The proliferating zone is again just behind the actual edge of the colony, and it is here also that calcification is initiated. Intercystidial septa arise when some of the ectodermal cells begin to secrete calcareous matter more rapidly than their neighbours. Such a septum divides by first forming a small swelling at its distal extremity; then secretion of calcareous matter from the ectodermal cells ceases in the centre of this swelling, but continues rapidly on either side.

In the Cyclostomata, in contrast to the Gymnolaemata, formation of a polypide precedes that of a cystid. Polypide rudiments arise in the growing zone, and Borg considered that it is the development of a polypide which initiates the fission of a septum necessary to create the two side walls of that particular zooid. The polypides develop in the usual bryozoan manner.

A new zooid is produced in the growth zone when the basal lamina gives rise to a septum which grows obliquely upwards to meet the terminal membrane (Fig.17C,H), thereby separating a new zooid from the growing edge below. The colonial form follows from the pattern of internal wall formation and the manner in which these walls merge and divide.

In *Crisia* the ancestrula is erect, and its two regions are separated by a joint. New longitudinal septa arise as growth proceeds, steadily creating new zooids (Fig.17B–D). In this and related genera the stem ramifies; additionally the branches are repeatedly divided by non-calcified joints into internodes of from one to many zooids. In most other cyclostomes the colony is basically adherent, although it may give rise to erect stems. The primary zooid is or becomes repent, with its basal side adnate to the substratum. New septa arise by fission of existing walls, but cystids will form only when newly produced septa fuse (Fig.17E, dotted lines), so that the zooid is free to grow upwards from the substratum and complete its development independently (Fig.17I).[130] The proliferating zone always remains at the edge of the basal lamina.

The young colonies of many cyclostomes pass through a stage identical to that of the young *Tubulipora* (Fig.17E,I). In this genus the colonies become lobate but generally remain adherent. In *Entalophora* the budding zone rises from the edge of the colony; the result is a stem growing upwards from an incrusting base. The adnate growing edge of *Berenicea* spreads so rapidly on each side of the primary zooid that the lobes meet, encircling the ancestrula, so creating a round colony with a peripheral growing edge.

Astogeny in Lichenopora

In *Lichenopora* and the closely related *Disporella* a circular colony is achieved in a different way. The primary zooid is funnel-shaped but, whereas in *Tubulipora* the proliferation of zooids occurs only on the adherent side, in *Lichenopora* budding takes place all around the periphery of the free end. As the accelerating production of zooids lengthens the circumference, so the basal lamina spreads back over the proancestrula (Fig.17J,K). Subsequent growth in *Lichenopora* involves processes, not found in most Recent cyclostomes, which are related to the coelocystic structure of the body wall.

Reconstruction of the early stages of cyclostome astogeny clearly shows how the stictocystic and coelocystic external wall structures have arisen. The formation of a primary zooid from the proancestrula proceeds similarly for both types (Fig.17F,G), as does the inception of the first dividing wall. What then matters is whether this wall (and all its successors) grows up to the terminal membrane and fuses with it (Fig.17H,I) or stops short of it so that adjacent zooids remain confluent (Fig.17J,K). In the stictocystic or *Tubulipora*-like condition, the cell layers bounding the new wall merge with those lining the terminal membrane; the cystids are then able to increase in length independently, and usually do so (Fig.17E,I), but new structures can be formed only inside the zooid lumen. With a coelocystic or *Lichenopora*-like condition the outer surface of the calcification is clothed by living cell layers, so that external secondary thickening can be deposited and various interzooidal structures created.

It has already been explained how a disciform *Lichenopora* colony forms. New zooids continue to arise at the periphery of the basal lamina by the fission of septa but, because the entire exposed surface is covered by living tissues, small coelomic spaces called *alveoli* are left between the zooids as they elongate (Fig.17K, al). The alveoli get larger as the zooics grow longer, and towards the centre of the colony—presumably to support the zooids—new walls arise and divide up the alveoli. Because of their mode of formation and site of origin, alveoli are generally not regarded as kenozooids or as zooid homologues at all.[119] Each alveolus may in time become partially closed by a terminal annulus (as in *Disporella*) or completely roofed over (as in *Lichenopora*). The lumen of the brood chamber, mentioned earlier (p. 106), consists of the united cavities of numerous roofed alveoli.

Astogeny in Heteropora

The colony of *Heteropora* starts to develop in the same way as that of *Lichenopora*, but three major differences become apparent as growth proceeds:[120] first, alveoli never form; second, most zooids, especially those situated near the centre of the colony, grow very long, making a stem; third, new zooids arise between existing zooids. These new, or interstitial zooids may remain small and lack polypides (kenozooids) or they may develop into autozooids. The formation of interstitial zooids is proof that the growing surface of the colony is covered by living cell layers. The brood chamber arises proximally as a fertile zooid, but subsequent breakdown of walls separating adjacent zooids produces a more extensive cavity in which the embryos develop.

Classification of Recent cyclostomes

The order Cyclostomata is of great antiquity, dating from the Ordovician at least. A satisfactory classification covering Palaeozoic, Mesozoic and Recent forms is impossible to achieve at present. The two most familiar groups today are the Tubuliporoidea and the Articuloidea, both with stictocystic walls. The former includes

Stomatopora, *Tubulipora*, *Entalophora*, *Berenicea* and *Diplosolen*. Some of these genera rank among the longest persisting in the animal kingdom, for both *Stomatopora* and *Berenicea* first appeared in the Ordovician. The Articuloidea, with jointed stems, are nearly as old, dating from the Silurian. *Crisia* is a large and well-known genus. Other superfamilies with living representatives usually placed in the order Cyclostomata are the Cancelloidea (e.g. *Hornera*), Cerioporoidea (e.g. *Heteropora*) and Rectanguloidea (e.g. *Disporella* and *Lichenopora*). All three have the coelocystic wall structure, and it seems possible that they should be removed from the Cyclostomata: this will be discussed later.

EXTINCT ORDERS

The Bryozoa have had a long and eventful fossil history. From the Lower Ordovician upwards a majority of limestone formations, especially those with shale alternations, are rich in bryozoan remains; they are less common in sandstone strata. The specimens themselves are generally calcified, but occasionally silicified and less easy to study, although the surface features may then be very clearly preserved.

Methods of study

The finest specimens are found in the shale layers associated with limestone, from which they may be recovered by carefully washing away debris.[3] Palaeozoic specimens can be freed from clayey particles by allowing pellets of caustic potash to deliquesce on the surface, following this treatment by thorough washing with water containing a trace of dilute acid, to neutralize any remaining alkali, and careful cleaning with a paint brush and very fine needles. A method used to recover silicified specimens from limestone is to dissolve away the latter with 3% hydrochloric acid.

Mesozoic and Caenozoic bryozoans can frequently be obtained from unconsolidated sediments, in which they may occur in vast numbers. Despite this abundance, the small size of the fragments results in their being easily overlooked, while their fragility necessitates careful treatment in the laboratory, soaking, washing and sorting the particles (see p. 142). The fragments of Palaeozoic species, on the other hand, are usually large enough to be readily visible to the naked eye. Where the fossils cannot be separated from the surrounding rock, as in some limestones, they may still be studied by sectioning. In many cases, in fact, particularly with the massive, solid or branched colonies of Palaeozoic bryozoans, specific identifications can be made only from carefully prepared sections which are thin enough to transmit light and show wall microstructure.

Preparatory methods are discussed by Ross and Ross, and other contributors, in the *Handbook of Paleontological Techniques*.[18] Pieces for sectioning must be cut with a rock saw or wire cutters and ground down on a coarse carborundum slab. Fragments are best handled by first embedding them in polyester resin and cutting with a special, thin-bladed saw. When a flat surface in the desired orientation has been obtained, it must be polished using a fine carborundum and finished off with an ultrafine water hone. The prepared surface is then cemented to a glass microscope slide with Canada balsam, using a hotplate or flame to dry and harden the balsam. To facilitate manipulation the slide is fitted into a cavity of appropriate dimensions prepared in a small wooden block, held in place by a film of water. The bryozoan fragment is then abraded on coarse, medium and fine wet carborundum blocks in turn or by a paste of graded carborundum powder on a sheet of plate glass. A thin section, which requires only a careful final polishing with a hone, will remain on the slide.

Fortunately it is now possible to prepare facsimile sections, adequate for many purposes, with a great deal less effort; although the genuine thin section remains unrivalled as a perfect specimen. To make a facsimile a polished surface must be prepared, as already described, which can be lightly etched to produce a mould from which detail can quickly be transferred to cellulose acetate film.[6]

The selected face of the specimen is honed and polished with stannic oxide powder. It is then etched for a few seconds with 5% formic acid, the exact time depending on the composition and texture of the material infilling the zooecia. The surface is washed and dried. It is then flooded with acetone and pressed lightly into pieces of cellulose acetate, 75×25 mm, cut from large sheets of thickness 1·5 mm. The preparation is allowed to dry, and the slide-sized piece of cellulose acetate is peeled off, washed and dried. Well-made slides display wall microstructure quite well and are invaluable for making routine identifications. The method facilitates study of Palaeozoic populations on a scale that would be unthinkable if wholly dependent on true sections.

A new approach, especially useful for studying fossil cheilostomes, makes use of internal moulds. The method, described by Hillmer,[1] is based on impregnation with polyester of the fine-grained sediments infilling fossilized skeletons and then dissolving away the walls with dilute hydrochloric acid. This method of preparing a mould to reveal the internal structure can be applied equally well to Recent material.

Palaeozoic Bryozoa

Although readily recognizable forms do not appear until the Lower Ordovician, the phylum must have arisen during the Cambrian if not before. Unfortunately, the non-calcified Phylactolaemata have left no fossil record; so that, while the evidence of comparative anatomy suggests that they are primitive, there is no geological evidence of their origin and evolution. Elsewhere in the phylum, on the other hand, there is an excellent fossil record covering four of the five generally recognized orders: Cyclostomata, Trepostomata, Cryptostomata and—from the time of their appearance in the Mesozoic—Cheilostomata. The remaining order, Ctenostomata, is also represented in Palaeozoic rocks, though very rarely as might be expected from the absence of calcification.

Quite recently some genera formerly regarded as cyclostomatous, together with some considered trepostomatous, have been grouped together as the new order Cystoporata.[116] This rearrangement has not gained universal acceptance[134] and the situation will be discussed later, after the trepostomes have been described.

TREPOSTOMATA

Characteristics

The trepostome zooids had calcified walls (zooecia) and produced substantial colonies of diverse form and size: many were massive, even exceeding 50 cm across. The size and abundance of the colonies, which sometimes contributed to coral-like reefs, make them important components of many Palaeozoic formations. Some species formed incrustations over other objects, in the manner characteristic of bryozoans in Recent times. The zooecia are usually more or less tubular and often of great length (Fig. 18A); their continued elongation produced hemispherical, conical or subglobular colonies or cylindrical outgrowths which divided into dendroid form or into a variety of foliaceous expansions.

The zooecia terminate at the surface of the colony as regularly arranged openings which have been termed *autopores* by some authors. Among these can be found small raised or depressed areas, or clusters of openings of a different size which stand out rather conspicuously. When these are elevated, forming mamillae or annuli, they are known as *monticuli*; when they form flat or depressed areas they are termed *maculae*. *Prasopora*, for example, had roundish colonies, the surface of which was freely covered with tubercle-like monticuli; *Constellaria*, one of the genera for which transference to the Cystoporata has been suggested, was a branched form in which the surface was marked with stellate maculae.

As in all bryozoa, the ultimate appearance of the colony is an attribute of the species; but shape is a labile character and a good deal of infra-specific variability occurred. Species cannot, therefore, be delimited solely on the gross characteristics of the colony, and it is necessary to consider the microstructure of the zooecia.

Trepostomes must be studied by means of thin sections, prepared as described earlier, cut in various planes. Longitudinal sections are most important, and they may be cut tangentially to the surface of the colony, thereby transecting the zooecia as they approach the periphery of the stem or branch, or they may be axial. An axial longitudinal section passes down the centre of the branch so that each of the radii displays the full length of a number of zooecia (Fig.18A). It is necessary to examine axial longitudinal and tangential sections in order to identify a trepostome species. Transverse sections through a branch may also be prepared.

Zooids, apart from the first in the colony, arose by budding from pre-existing zooids; they developed as long, straight or curved tubes, circular or polygonal in section, opening to the exterior by a simple terminal orifice. Very noticeably in trepostomes the zooecial walls are thin in their axial or oldest part whereas, towards the periphery of the colony, for the final fraction of their length, they become thick with lamellar calcification (Fig.18B,D). It is this change in the nature of the zooecium which is commemorated in the name Trepostomata (*trepos*, change+*stoma*, mouth). In erect colonies this differentiation between inner and outer parts of the zooecia gives rise to corresponding axial and peripheral zones in the stems (Fig.18A).

Fig. 18 Zooecial structure in the Trepostomata

A Part of branch in longitudinal section
B Distal part of single zooecium in longitudinal section showing amalgamate appearance
C Zooecium B in tangental section
D Distal part of single zooecium in longitudinal section showing integrate appearance, together with an acanthopore
E Zooecium D in tangental section
F Part of zooecium with cystiphragms
G Arrangement of first zooecia in a colony (*A* ancestrula; 1 first generation buds; a2, p2 anterior and posterior second generation buds)
H Part of zooecium with mesopore (on left)

(A, B, C based on Boardman;[118] D, E based on Boardman[118] and Tavener-Smith;[136] F based on Cumings and Galloway;[127] G based on Cumings[126]).

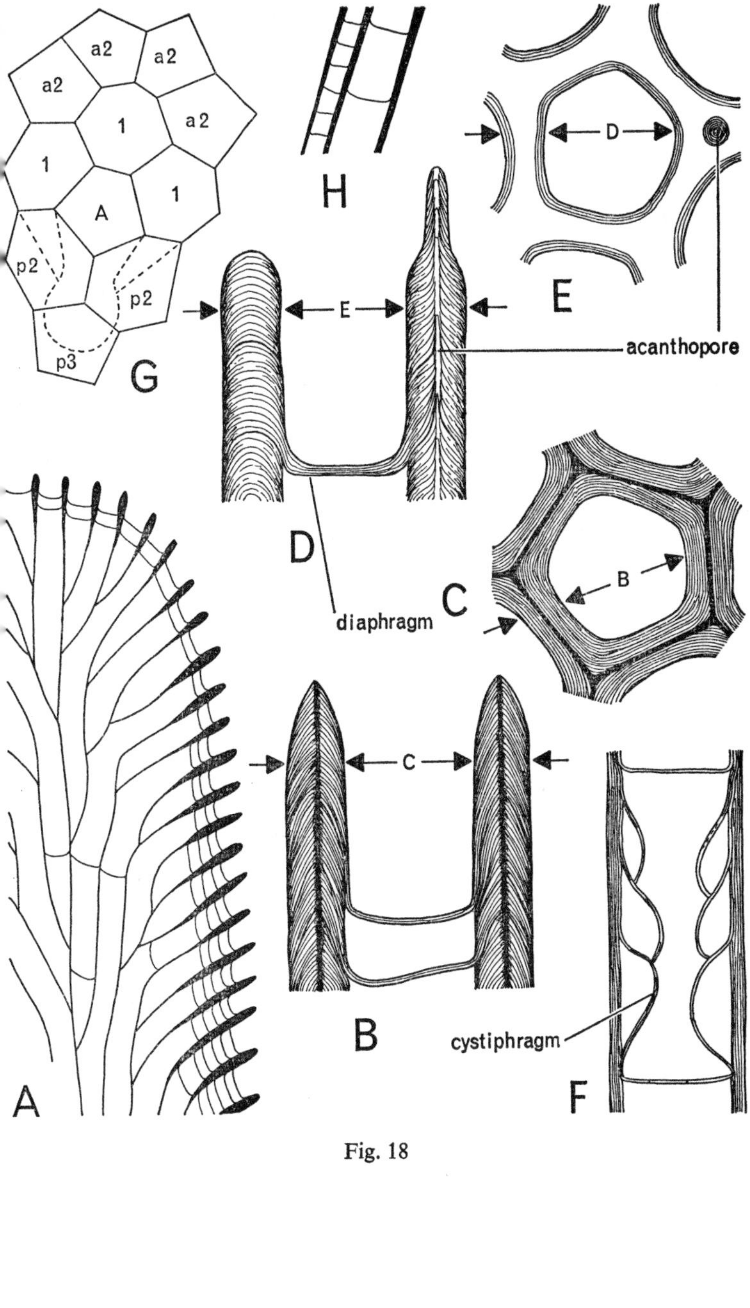
a2
a2
a2
a2
1
1
1
A
p2
p2
p3
G
H
D
E
E
acanthopore
D
diaphragm
C
B
C
B
cystiphragm
F
A

Fig. 18

The walls display a laminated microstructure. When viewed in tangential section the shared walls sometimes show a distinct dark line (Fig.18C), apparently separating the two components of a duplex structure. Such walls were termed *integrate*, and it was once thought that the dark line represented the primary zooecial boundary upon which further calcification was deposited by the epithelia of the contiguous zooids. Other walls show a thicker, irregular, non-laminate central zone (Fig.18E) and were termed *amalgamate* in the belief that the boundary had become obscured by the union of the two wall components. The two types of structure were regarded as fundamental and made the point of distinction between the sub-orders INTEGRATA and AMALGAMATA, divisions of the Trepostomata which persist even in some recently published texts, although practising bryozoologists have not used them in this way for many years.

The careful morphological studies of Cumings and Galloway[127] showed that the trepostome wall is never truly duplex. Axial longitudinal sections revealed that sometimes the growing edge of the inter-zooecial wall tapers so that the laminae comprising the calcification converge to the apex at an acute angle: the series of abrupt flexions made by the laminae at their apices produces the thin dark line seen in sections (Fig.18B). Other walls have a rounded growing edge and the laminae curve steadily across the wall from one zooecial lumen to the other (Fig.18D). The region of greatest curvature in the centre of the wall may appear darker (or lighter), but this arrangement never gives rise to a sharp thin line. Amalgamate walls seen in tangential section appear to be laminated only in a zone adjacent to the zooecial lumen. The reason is, of course, that in the centre of the wall the plane of the section runs parallel to, and not across, the laminae (Fig.18D,E).

Recent work by Boardman,[118] who provides a valuable account of the Trepostomata, confirms the findings of Cumings and Galloway, so that it is quite impossible to regard the two types of wall structure as bases for the maintenance of suborders. Both amalgamate and integrate structure may be found in colonies of the same species, in the walls of a single colony or even within the length of one zooecium.

At intervals the zooecia are crossed by partitions or *diaphragms*, which are well spaced out in the axial region but rather close together near the surface (Fig.18A) or, as in *Hallopora*, in the inner part of the peripheral zone. Diaphragms may form complete partitions, they may be centrally perforate or they may be otherwise incomplete, reaching only part way across the zooecium. Diaphragms also show a laminated structure and flex orally as they meet and merge with the zooecial boundary wall (Fig.18B,D). Another type of structure is the *cystiphragm*, which is a more or less annular septum inside the zooecium. Its appearance suggests that the secretory epithelium blistered away from the wall and deposited a thin calcareous layer in its new position (Fig.18F). Cystiphragms clearly reduced the volume of the zooecial lumen, but did not very obviously perform any other function.

The autozooecial orifices or autopores are frequently interspersed with smaller openings termed *mesopores* (Fig.18H). Their presence suggests that, as in extant orders, polymorphism of zooids occurred in the Trepostomata (although some of these specialized 'pores' may be interzooecial). Very small specialized zooids occur among the autozooids in both cyclostomes and cheilostomes. These may contain polypide rudiments, like the nanozooids of *Diplosolen*, or they may be kenozooids. The mesopores of trepostomes are usually regarded as kenozooecia. The orifices of incipient autozooecia (see below) are also small, and can be mistaken for true mesopores.

Sections through the peripheral region of trepostomes frequently reveal rodlike or tubelike structures of various kinds, known as *acanthopores* (*acantha*, thorn), which penetrate the zooecial walls and may be manifested at the surface as a dot, pustule or spine. The most distinctive and typical acanthopore structure consists of a clear calcite axis with walls of steeply inclined laminae (Fig.18D,E), which extends for a considerable distance through the peripheral region of the zooecial wall.[133]

Tavener-Smith[136] has made a study of the acanthopores of *Leioclema* from the Wenlock Limestone (Silurian). Each arises at a point-locus of accelerated growth within a zooecial wall, the laminated nature of which is repeated in the characteristic cone-in-cone structure of the acanthopore. The 'pore' is not hollow at all, and the appearance of a lumen is an illusion, resulting from the contrast of many thin laminae in the longitudinal plane against the central zone in which the laminae become much wider as they approach the apex of each cone. In *Leioclema* at least, then, the acanthopore is not a tube and cannot be regarded as a heterozooecium. The acanthopore protrudes beyond the orifice of the autozooecium as a spine (Fig.18D). The function of such processes may simply have been to provide a bristly surface likely to discourage larvae seeking a place on which to settle or small marauding predators ready to devour the polypides; alternatively they may have supported the cuticle and eustegal cell layers of the colonial wall.

Cystopores (discussed fully on p. 120), which may represent interzooecial strengthening, occur in certain trepostomes, e.g. *Leioclema* and *Constellaria*.

Special brood chambers have not been described for trepostomes. The clustered pores comprising maculae are frequently of greater diameter than those of autozooecia. They belong to *mega-* or *macro-zooecia* which may have been the fertile or reproductive zooids of the colony.

Astogeny

The Trepostomata were established as a bryozoan order by Ulrich in 1882, but the view that at least some of the trepostome families were tabulate corals persisted for some years. Proof of their proper identity was provided by Cumings,[126] who studied the early astogeny in a number of genera. *Prasopora* is a typical example.

The proancestrula (Fig.18G) is a circular disc hardly demarcated from the tubular portion of the primary zooecium (A). The wall of the proancestrula differs distinctly in texture from those subsequently formed, being clear and lacking fibrous or lamellar structure. The tubular region is at first prone, but soon curves upwards from the substratum. The first daughter zooecia (Fig.18G,1) arise from the distal side of the primary zooecium. The first posteriorly directed zooecia (p2), like the next generation of anteriorly directed zooecia (a2), do not arise directly from the primary zooid.

The presence together of the proancestrula and a mode of budding almost identical with that found in cyclostomes confirm the bryozoan nature of the trepostomes beyond reasonable doubt. In corals development is direct from the planula and the buds arise symmetrically around the first zooid.

The simplest colony form,[122] and the precursor of more complex shapes, is a crust attached to the substratum by a basal lamina. The zooecia radiate from the point of origin of the colony rather as in a cyclostome, and are at first adherent, contributing to the basal lamina. As they increase in length, the zooecia curve upwards to open at the surface of the colony. The circumference of the basal lamina at this stage constituted the growing edge and budding zone where new zooecia arose by the fission of existing walls.

In colonies which were hemispherical in shape, formation of new zooecia was not confined to the periphery of the basal lamina but could occur all over the exposed surface, as in the Recent genus *Heteropora* (p. 110). In an upwards growing colony this surface became the principal instigator of new zooecia, and must have been covered with living tissues as in *Heteropora*. Since the lengthening zooecia must have diverged radially as the domed growing surface of the colony spread, spaces would have been left unless new zooecia filled the gaps. Often such infilling zooecia were kenozooecia, which would have lacked polypides and contained frequent diaphragms, and clearly had a supporting function. Hemispherical colonies need not display distinct central and peripheral zones, though the zooecia themselves may have done so, because new zooecia were intercalated all over the surface at ever-increasing distance from the colony primordium.

The colonies were very frequently lamellate, with layers of growth superimposed upon each other. Clear signs of detritus may be discernible between the layers and it looks as though, for some reason, growth ceased, only to continue again after a period of quiescence. Borg's[122] studies suggested that both auto- and kenozooecia could be closed off by diaphragms at the surface, and that this certainly happened when one of these periodic cessations of growth took place.

When an incrusting base gave rise to an erect stem, the zooecia at the point of eruption first increased in length and later changed their direction of growth. Rising upwards from the crust, they then curved away from one another, producing new zooids medially so that no gaps were left (Fig.18A). It is in such stems that strongly differentiated axial and peripheral zones are most apparent. The rate of elongation of individual zooecia must have declined as their orientation became increasingly horizontal, otherwise a stem-like structure could not have resulted. The thicker walls and closer diaphragms seen in the distal parts of the zooecia are thus a reflection of the retarded growth rate at the periphery of the stem.

Diaphragms seem generally to be sparsely developed in the axial zone and, of course, in the young cystids at the stem apices, but become more numerous in the peripheral zone, dividing the zooecia into small chambers. The orally directed flexures of the diaphragms indicate that the secretory epithelium was on their distal side. The polypide in each zooecium must have degenerated whenever a diaphragm was deposited, a new polypide later forming beyond the diaphragm. Fossilized remains have been discovered in the closed chambers of the zooecia which have been interpreted as brown bodies—the semi-permanent residue resulting from the degeneration of the polypide within the cystid in Recent bryozoa (p. 59).

In the peripheral zone the zooecial chambers become shorter (Fig.18A), although it must be realized that the interval between successive diaphragms does not equal the length of a chamber when it was terminal and contained a polypide. A zooecium here has the form of a nesting stack of tumblers. The living tissues covered the whole surface of the colony, with polypides confined to the distalmost chamber of each zooid; the whole structural mass axial to the distalmost chamber can have served no function other than support.

CYSTOPORATA

Characteristics

Two apparently related groups of Palaeozoic bryozoans, Ceramoporoidea and Fistuliporoidea, have been in the past variously placed in the Trepostomata or the Cyclostomata. Their characters, however, do not accord fully with those of either of these long-established orders, and a recent proposal[116] unites the two superfamilies as the principal components of a new order, Cystoporata.

The trepostomes had tubular, wholly contiguous zooecia, with more or less polygonal orifices and numerous diaphragms whose flexures contributed to a layered wall, much thickened peripherally. Mural pores were absent. The colonies were further characterized by the occurrence of groups of larger zooecia forming maculae and by mesozooecia and acanthopores. Most Palaeozoic cyclostomes belonged to one or other of the still extant groups Tubuliporoidea or Articuloidea, the characters of which are well known. The remaining superfamily, Hederelloidea, was of the same basic type. The zooecia in all three groups are tubular, basally tapering, often free distally and terminating in a circular orifice. Their walls are partly prismatic and partly laminar, but not heavily layered, and contain interzooecial pores.

Zooecia in the Cystoporata arise rather abruptly from a flat basal lamina, they remain adjacent throughout their length, and they contain frequent diaphragms (Fig.19A,B); in these respects, and in the possession of maculae, the cystoporates resemble trepostomes. The zooecial walls, however, have a granular-lamellar microstructure more like that of cyclostomes, and mural pores are present. Other characteristics differentiate them from both orders: the orifices are often bounded by a crescentic thickening called a *lunarium*; and, secondly, variant 'pores' are present having the nature of thin-walled tubes divided up by sparse or abundant partitions. These have been named—in an extension of an unsatisfactory terminology—*cystopores* (hence Cystoporata). They are probably interzooecial structures,[133] and not mesozooecia.

When erecting the order Cystoporata, Astrova[116] also included in it two trepostome families, Constellariidae and Dianulitidae; but this regrouping has elicited criticism.[134] Since the affinities of these two families cause controversy among specialists, in this book they will be left in the Trepostomata, and the Cystoporata will be maintained for the Ceramoporoidea plus Fistuliporoidea.

Cystoporate colonies grew in many shapes: rounded, massive, branched or bilaminate. There is frequently differentiation into axial and peripheral zones. The cystopores may open to the surface as maculae. In the Ceramoporoidea (Fig.19A) neither the cystopores themselves nor the partitions within them are plentiful, whereas in the Fistuliporoidea (Fig.19B) the cystopores are numerous, contain abundant partitions and are roofed at the surface.

Astogeny

The Swedish zoologist F. Borg[122] concluded a lifetime of studies on Recent cyclostomes by turning his attention to Palaeozoic fossils, including the Ceramoporoidea and Fistuliporoidea, paying careful attention to the mode of colony development. Borg's manuscript was unfortunately unfinished at his death, and in its published form it is difficult to pick out clearly all the conclusions its author may have intended. Nevertheless, his main findings are clear.

The proancestrula and the tubular part of the primary zooecium are not sharply set apart, suggesting a primitive condition with an undifferentiated ancestrula. The young colonies are cornet-shaped and became disciform by overgrowing the proancestrula. They adhered to the substratum by a basal lamina made up of zooecial walls as in trepostomes and cyclostomes, and the formation of new zooecia took place at the periphery of the lamina only. Later astogeny shows clearly that the walls were coelocystic, as in trepostomes but unlike Palaeozoic cyclostomes.

In the Ceramoporoidea (*Ceramopora*, *Ceramoporella*), there was a tendency for zooecia to arise in linear series, forming more or less distinct radiating lines of zooecia which alternated with interradial depressed areas between them. These areas became bridged by a series of transverse bars, evidently to buttress the distal parts of the zooecia. Mural pores were present between zooecia, and between zooecia and the interradial areas.

In *Fistulipora*, away from the colony margin, the zooecia were also free distally. In the interzooecial spaces new walls were laid down, creating a system comparable, in Borg's view, to the alveoli found in the Recent genus *Lichenopora* (p. 109). The alveoli became roofed over, and above them secondary and even tertiary alveoli developed. The cystopore maculae of Astrova are, of course, the alveoli of Borg. It appeared to Borg that the interradial grooves in the Ceramoporoidea represent a somewhat simpler evolutionary stage than that seen in the alveoli of the Fistuliporoidea. It was the extrazooecial nature of these structures that led Borg to conclude that the colony must have been covered by colonial epithelia, exactly as in *Licheno-*

pora. The presence of this layer would also explain easily the development of lunaria.

No enlarged brood chambers have been recognized; but in *Fistulipora* Borg found that some of the older zooids were dilated distally and partly overgrown by alveoli. This suggested to him the situation from which the lichenoporine brood chamber could have evolved.

In some species of *Fistulipora*, just as in *Lichenopora*, multiple colonies were found. This condition can be explained by assuming a cessation of growth, perhaps of a seasonal nature, at the peripheral budding zone. When the proliferation of zooids recommenced it did so in a discontinuous manner creating a lobate and ultimately multicentred colony. In the related genus *Cyclopora* erect stems occurred. It might be supposed that these would have arisen from the centre of the colony (as in the Recent *Heteropora*), but this cannot have been so. Since the lunarium was borne on that side of the orifice nearest to the centre of the colony, on any centrally rising stem the lunaria would inevitably be present on the upper side of the orifices: in fact they occurred on the lower side. Evidently an indentation formed at the growing edge, perhaps through meeting an algal stem. When the two lobes so formed rejoined they left behind a small lacuna of growing edge which could only grow upwards, incrusting the stem or perhaps as a free hollow tube.

Borg concluded—as has generally been recognized—that the Ceramoporoidea and Fistuliporoidea were fairly closely related, perhaps with a common ancestor. The former apparently display the simpler kind of alveolar structure and, consistent with this view, flourished in the Ordovician and Silurian, whereas the Fistuliporoidea reached their peak abundance between the mid-Silurian and the mid-Carboniferous (Fig.20).[116]

Borg[122] also studied the Ordovician–Silurian genus *Spatiopora*, supposedly referable to the Ceramoporoidea,[3] but possessing many trepostome characters, including the budding of zooecia away from the colony margin. Its taxonomic position needs re-assessment.

CRYPTOSTOMATA

The second long-established order of extinct and almost exclusively Palaeozoic bryozoans is the Cryptostomata, introduced by Vine in 1883. The zooecium is basically of the tubular kind characteristic of other stenolaemates; indeed the boundary between the trepostomes and cryptostomes is not always clearly defined.[133] The distal part of each zooecium is clearly differentiated from the proximal part, and s termed the *vestibulum*; it is frequently delimited proximally by a shelf or *hemiseptum* on one side of the tube, or there may be two hemisepta. It has sometimes been supposed—though apparently without evidence—that the hemiseptum marks the position of the true orifice (hence Cryptostomata, from *kryptos*, hidden+*stoma*, mouth) and may have supported some kind of operculum. The very length of the vestibulum in certain cryptostomes makes this seem rather improbable, as it would demand the presence of a very long introvert (tentacle sheath) and a remarkably efficient hydrostatic protrusion mechanism. It is perhaps more likely that the hemiseptum was in some way involved with the mechanism of tentacle eversion, such as providing a point of origin for muscles. The zooecial orifice may be surrounded by a narrow rim or peristome, as in *Sulcoretepora*. Within the peristome may be a *lunarium*.

Acanthopores occur, as in the Trepostomata, and may be large (*megacanthopores*) or small (*micracanthopores*). Mesopores may also be present.

The colonies of cryptostomes are characterized by a delicacy lacking in trepostomes. Especially notable are the erect funnel- or fan-shaped colonies of the fenestelloids with their wide, reticulate fronds (Fig.19D). The Russian workers Astrova and Morozova[117] have proposed a classification of cryptostomes based on the structure of the colony, and recognize three superfamilies: PTILODICTYOIDEA, RHABDOMESOIDEA and FENESTELLOIDEA. The first two of these are the more closely similar, having branching colonies and tubular zooecia. In the Ptilodictyoidea (e.g. *Sulcoretepora*) the fronds are flat and have a bilamellate construction in which the zooecia arise from the two sides of a *median lamina* or mesotheca—equivalent to the adpressed basal laminae of the two layers—and open on their respective surfaces (Fig.19C). In the Rhabdomesoidea (e.g. *Rhombopora*) colonies comprise a bifurcating arrangement of slender stems and branches, in which the zooecia radiate from a central axis.

When zooecia arise from a median lamina or axis, they at first grow parallel to it as thin-walled tubes, but then curve rather sharply to the surface through a zone of extensively developed extrazooecial calcification. A sectioned colony thus displays a lightly calcified axial zone and a heavily calcified peripheral zone as in trepostomes, but the boundary is here more strongly marked (Fig.19C); and the thick calcification has a different appearance in the two orders. In those ramose cryptostomes with relatively long zooecia, diaphragms may be present as in trepostomes; but the thin fronds and delicate form which characterize the Fenestelloidea are obviously incompatible with the formation of very long tubes, and the zooecia are here very much shorter. (In the primitive fenestelloid genus *Archaeofenestella*, however, diaphragms are present in the zooecial chambers.)[132] Interzooecial pores are not present in cryptostomes.

In the Fenestelloidea, the colonies are reticulate and unilamellar, and the zooecia characteristically flask-shaped. There is also a distinctive pattern of calcification, described below, and a budding method which differs rather greatly from that prevailing in other stenolaemates. Elias and Condra[128] have already proposed that, on the basis of their general structure, the fenestelloids deserve ordinal status as the Fenestrata. If the presence of a distinctive budding system be confirmed, the proposal might seem not merely logical but necessary.

Fenestelloidea

The fenestelloid frond usually arises from the supporting base as a fan or funnel, but in *Archimedes* the lamina is disposed in a continuous spiral around a central axis. Zooecia are arranged in series along the numerous, regularly spaced, gradually diverging and occasionally bifurcating rami which comprise the frond and link up with each other at regular intervals by means of *crossbars* (also known—though not very appropriately—as dissepiments) generally devoid of zooecia (Fig.19D). (In the Ordovician phylloporinids, possibly precursors of the later fenestelloids, crossbars are not differentiated and the zooecia extend along all the anastomosing branches.) The zooecial orifices all open on the same (frontal or obverse) side, which usually bears a median longitudinal ridge or *carina* ornamented by spikes or *nodes*. The frontal may be the inside or the outside of the funnel according to genus. In *Fenestella* the zooecia form two series in each ramus, but the rows are more numerous (3–7) in *Polypora*.

Fig. 19 Structure in the Cystoporata and Cryptostomata

A Zooecia and cystopores of *Ceramopora*
B Zooecia and cystopores of *Fistulipora*
C Longitudinal section through part of frond of a bilaminate cryptostome (Ptilodictyoidea)
D Part of fenestelloid frond from the growing edge in frontal view
E Transverse section through the branch of a fenestelloid
F Reconstructed longitudinal section through the apex of a fenestelloid branch showing the method of budding. Stages in the formation of a cross-wall are numbered 1–5

(A, B after Astrova;[116] E after Tavener-Smith;[135] F based on Tavener-Smith[135]).

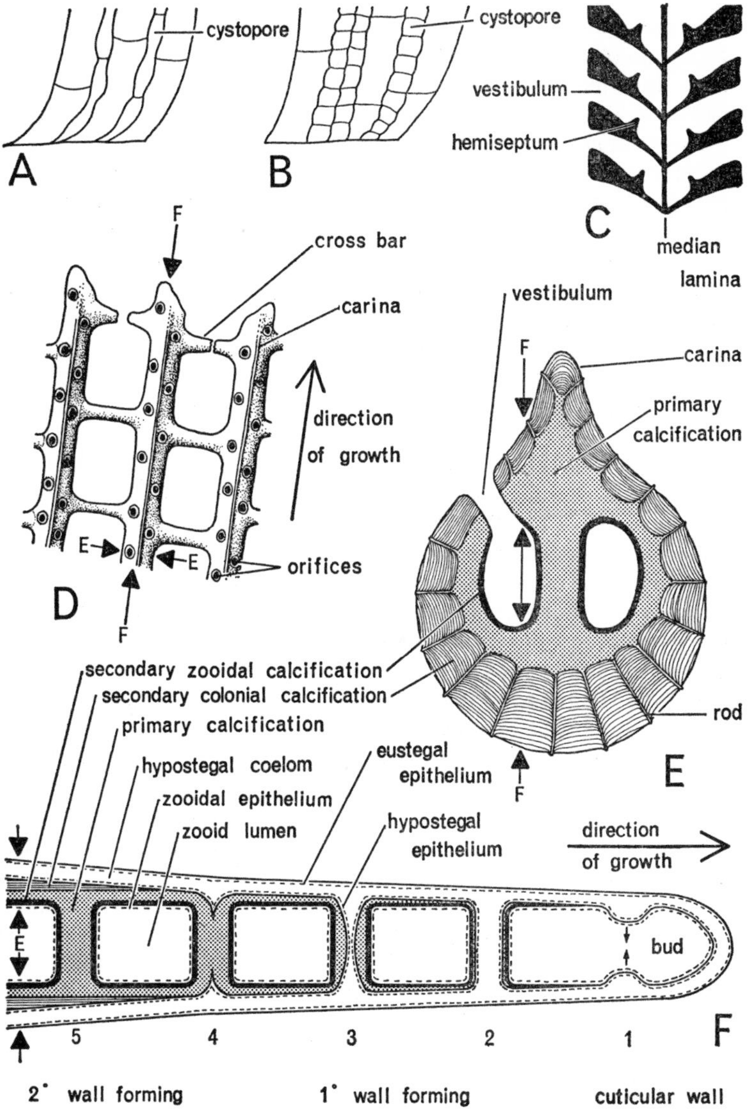

cystopore
A
cystopore
B
vestibulum
hemiseptum
C
median
lamina
F
cross bar
carina
direction
of growth
E
E
orifices
F
D
vestibulum
F
carina
primary
calcification
secondary zooidal calcification
secondary colonial calcification
rod
F
E
primary calcification
hypostegal coelom
zooidal epithelium
zooid lumen
eustegal
epithelium
hypostegal
epithelium
direction
of growth
E
bud
5
4
3
2
1
F
2° wall forming
1° wall forming
cuticular wall

Fig. 19

Sections through the rami reveal three distinct skeletal regions (Fig.19E): first there is a very thin, laminated layer which lines each zooecium (shown black in the figure); then there is a central granular zone in which the zooecia—apart from their vestibula—appear embedded (shown stippled); and finally, completely surrounding the granular zone, is another laminated deposit, thick and traversed by radiating rods. The granular zone, formerly known as 'germinal plate' or 'colonial plexus', has been revealed as primary calcification by Tavener-Smith's fine study of *Fenestella*;[135] while the laminated portions were secondarily deposited.

The zonal construction of the rami, together with the uninterrupted nature of all but the inner calcification, make it evident that the skeleton comprises the depositions of two epithelia: zooidal and colonial. Furthermore, the structure is so similar to that described by Borg[119] for the extant cyclostome genus *Hornera* (p. 105), that the disposition of living tissues seems likely to have been the same. This assumption—in effect that the fenestelloid wall was a coelocyst throughout—has permitted Tavener-Smith[135] to suggest a method of frond extension, together with the concomitant processes of zooidal budding and skeletal deposition, which, for the first time, satisfactorily explains all the observable anatomical features of the frond (Fig.19F).

There is a sharp discontinuity between the laminated layer lining the zooecia and the granular, primary skeleton; whereas the latter merges gradually into the outermost zone. It may be concluded, therefore, that the primary calcification was colonial and not zooidal, and was secreted by the hypostegal epithelium of the coelocyst. Deposition was presumably continuous at first, since the primary skeletal matter is non-lamellar, and each newly formed cystid was rapidly encased as soon as it had attained full size. Later, as the rate of secretion decreased and became intermittent, the laminated calcite which characterizes the thick, outermost zone of each ramus was produced. For some reason, however, at a number of point-loci, the hypostegal epithelium must have continued skeletal deposition in the original manner; so, series of rods corresponding to these loci now apparently radiate from the primary layer and traverse the outer zone, ending up as pustules at the surface. Possibly, at the point of formation, the rod tips performed the function of supporting or anchoring the eustegal epidermis and cuticle, as was suggested for the acanthopores of trepostomes.

The carinae, nodes, crossbars and other components serving to support the colony were likewise all products of the hypostegal epithelium. The laminated layer lining each zooecium was, on the other hand, secreted by the zooidal epithelium.

Fenestelloid budding

Budding of zooids, Tavener-Smith's studies suggest, took place at the frond extremities in a manner analogous to that now found in the Gymnolaemata: namely by the separation of a chamber on the distal side of the zooid by an ingrowing annular wall (Fig.19F). The exact method by which partition was accomplished, however, must have differed fundamentally from that which typifies the Gymnolaemata (Chapter 3, p. 57; Chapter 5, p. 86), for each successive cystid was not split off by a perforate septum but by a double wall which was entire before any calcification took place. The daughter cystid, therefore, was nourished via the hypostegal coelom which enveloped it, and not by direct communication with the parent zooid.

Since the interzooecial walls contain primary calcification, the hypostegal epithelium as well as the zooidal epithelium must have participated in the ring-like invagination which constricted off the bud (Fig.19F, position 1). Calcification of appreciable thickness would have been possible only if the infolding hypostegal epithelium carried within it an extension of the hypostegal coelom, so that the newly formed wall was hollow (Fig.19F,2). Subsequently, each of the now separate parts the hypostegal epithelium formed, when the annular invagination had unified, started to deposit calcite in the usual way between themselves and the zooidal epithelium (Fig.19F,3): they thus inevitably approached each other and finally fused at about the midpoint of the wall. The re-united hypostegal epithelium withdrew peripherally in the manner of an iris diaphragm being opened (Fig.19F,4), still secreting calcite: the wall was complete (Fig.19F,5).

This mode of zooid formation appears, at first sight, far removed from that of typical stenolaemates from which it must have evolved. A little thought shows that this is not really the case, and the fenestelloid condition is readily derivable from a unilamellar frond which budded zooids from a basal lamina by the fission of septa. A basal lamina can never have a coelocystic structure so long as it is adnate (see Fig.17K) but, as soon as a lobe rises from the substratum, the epithelia can spread from the growing edge down the basal surface like an external mantle. An immediate advantage would be the ability to reinforce the basal lamina by secondary calcification, so strengthening the free-standing frond.

Few further changes would be required. The first, perhaps, was a slight delay in the onset of calcification from the zooidal epithelium behind the growing edge. Since the zooidal and hypostegal epithelia would then be closely approximated, the delay would facilitate involvement of the latter in the invaginating crosswall. An extended interface between the hypostegal coelom and the zooid bud would presumably have ensured an improved nutrient and oxygen supply to the bud. Thereafter the hypostegal epithelium must gradually have taken over the role of laying down the primary skeleton until the fenestelloid condition was reached. The obvious advantage is that a thick interzooidal wall, which braces the young frond, could be deposited without simultaneously decreasing the volume of the cystid.

Fenestelloid astogeny

The general similarity in colony form between the fenestelloid cryptostomes (Fig.19D) and the reteporine cheilostomes (Fig.1I) inevitably provokes comparison. Both types presumably evolved in response to identical environmental pressures: the need for the filtration area of the colony to be as large as possible in habitats not subject to appreciable water movement. However, neither the reticulate structure itself nor the method by which it is achieved are alike. Cumings,[124,125] in his study of the early astogeny of fenestelloids, seems to have regarded a resemblance to the budding pattern of the retepore *Iodictyum* as evidence for a similar budding method. Nevertheless, his two papers are important, especially in trying to elucidate the systematic position of the Cryptostomata. He studied a mid-Devonian species now thought to have belonged to *Fenestrapora*.

The early colonial development is obviously highly adapted to the role of providing a supporting stem for the fronds; and the pattern is, moreover, greatly obscured by the subsequent deposition of secondary calcification. The primary zooid of the fenestelloid apparently corresponds mainly to the proancestrula of trepostomes and cyclostomes, for the tubular part is short. Tavener-Smith believes that the colonial epithelia must have overgrown the ancestrular wall before the first bud could have been produced. Astogeny proceeds[124,125] by the production of a circle of centrifugally directed adnate zooecia which constitute the first part of the basal plate of the colony. Tiers of zooecia are added above the plate, in columnar fashion at first but later diverging in a funnel-like manner. Primary carinae or supporting ribs grow upwards from the secondary calcification of the basal plate.

Cumings found that the zooecia in the first formed or neanic (p. 56) part of the colony are tubular, and that the characteristic ephebic or vestibulate configuration did not appear until the frond started to form. He stressed the close resemblance between the fenestelloid and cyclostome proancestrulae and between their aeanic zooecia. The similarity in early astogeny between fenestelloids and trepostomes is also marked, except for the manner in which the tem is built up. Although frond form and zooecial shape in the phebic part of the fenestelloid colony have evolved *pari passu* in heir own individual manner, the early astogeny provides clear vidence of the stenolaemate affinities.

7

FOSSIL HISTORY AND GROUP INTERRELATIONSHIPS

STENOLAEMATA

Cystoporata, Trepostomata and Cryptostomata

A poorly preserved fossil called *Archaeotrypa* from the Uppe Cambrian may perhaps be a primitive ceramoporoid, and th earliest known bryozoon. Definite ceramoporoids have been reporte from the Lower Ordovician of arctic Russia and of the Unite States.[134] It seems that the ceramoporoids were the first stenolaemate to appear, although the trepostomes and cryptostomes soon followe By the Upper Ordovician the major groups were all well establishe including the fenestelloids (Cryptostomata) and tubuliporoid (Cyclostomata).[19] The Ceramoporoidea survived until the Devonian but the Fistuliporoidea, which seem to be a later offshoot from th same stock, continued with fair success throughout the second ha of the Palaeozoic.

The dominant bryozoa of the later Lower Palaeozoic were th Trepostomata. From the time of their appearance in the Lowe Ordovician they evolved with rapidity to reach their apogee durin the upper part of the same period. From the end of the Ordovicia the trepostomes started their long decline, although they remaine important until late in the Palaeozoic.

Compared to the trepostomes, the geological life span of th cryptostomes presents a distinct contrast. They, too, evolved rapidl during the Ordovician, without quite matching the early success o the trepostomes. Thereafter, the cryptostomes continued to flouris for millions of years and were the dominant bryozoa during th Devonian, Carboniferous and Permian periods. Their demise wa sudden for, during the Permian period, the population crashed and the order was practically extinct by the close of the Palaeozoic.

The success of a fossil group may be gauged in part from a measure of diversity, such as the number of genera, at given intervals. Other instructive parameters, however, are rates of appearance and extinction of genera, calculated in numbers per million years (m y). Thus as evidence for a rapid initial rate of evolution in the Trepostomata, it may be noted that the rate of appearance of genera

Period	Duration	Trepostomata			Cryptostomata		
	(m y)	Genera	Firsts	Lasts	Genera	Firsts	Lasts
Permian	45	4	0·4	0·9	36	1·6	8·0
Upper Carboniferous	35	8	0·8	1·1	33	2·9	0·9
Lower Carboniferous	45	17	2·9	2·7	43	6·0	4·2
Devonian	50	22	3·2	4·2	51	7·0	7·4
Silurian	40	14	0·5	1·3	30	4·3	3·5
Ordovician	60	50	8·3	6·7	49	8·2	6·0

Geological history of the Trepostomata and Cryptostomata, showing the number of genera recorded from each period and the rate of first and last appearances of genera (number per 10 m y) during each period of the Palaeozoic era.

during the Ordovician, although very high (see accompanying table), was practically equalled by the rate of their extinction. Evolutionary and extinction rates of both trepostomes and cryptostomes fell sharply during the Silurian. This period was, therefore, either one of little evolutionary change or just manifestly understudied,[5] or perhaps both.

For trepostomes both rates apparently recovered somewhat during the Devonian, but declined steadily thereafter. Never, after the close of the Ordovician, did the rate of formation of new genera noticeably exceed the rate of their extinction. In cryptostomes and fistuliporoids, on the other hand, the evolutionary rate reached a second peak during the Devonian and Lower Carboniferous. The decline of both of these groups was characterized by a sharply falling evolutionary rate during the Upper Carboniferous and Permian and finally, in sharp contrast to the pattern noted in trepostomes, a soaring extinction rate.

The close of the Palaeozoic marks the end of a great era in the geological history of the Bryozoa. Trepostomes, fistuliporoids and cryptostomes all disappeared in the Permo-Triassic wave of extinctions. With them went many other major marine animal groups, including eurypterids, trilobites, blastoids and certain lines of brachiopods and crinoids. Yet, as Rhodes[22] has pointed out, the Permian as a whole was not characterized by a gradual dwindling of those stocks that became extinct towards the close of the period. The difference in pattern of decline between trepostomes and cryptostomes (Fig.20) makes a common cause unlikely; although in both, as in many of the other groups that died out, the rate of production of new genera first started to fall markedly during the Upper Carboniferous.

Suggested causes for Permo-Triassic extinction are many. They include: climatic changes, for while the northern land masses experienced desert conditions, the southern continent (Gondwanaland) was affected by a long, fluctuating ice age; a major decline in plankton production; affects of the Variscan orogenesis; eustatic changes in sea level which eliminated the continental shelf; and, in the realm of pure hypothesis, that the declining groups had simultaneously all developed evolutionary senility. Rhodes concluded that, since Permo-Triassic extinction was not one catastrophic event, a single cause was unlikely. Moreover, the general absence of Permo-Triassic transitional marine sequences has probably exaggerated the sharpness of the faunal break.

The decline of the trepostomes might have been an effect of competition with the more successful cryptostomes. The sudden demise of the latter is, however, a genuine example of Permo-Triassic extinction, for bryozoa have not been reported in any numbers from Triassic marine deposits.[5,11] Since cryptostomes were —like Recent bryozoa—predominantly shelf forms,[11] any regression of the shallow sea would obviously have had a deleterious effect upon their numbers; although other factors may well have been involved.

Cryptostome ancestry for the Cheilostomata?

There is striking convergence of both colonial and zooecial form between the Cryptostomata (particularly the fenestelloids) and the Cheilostomata. This has led to a theory, notable mainly for its circular argument, that the cheilostomes evolved from the cryptostomes. The presence of a 'hidden orifice' in cryptostomes—situated at the level of the hemiseptum—has been inferred, not on definite structural evidence, but from analogy with certain cheilostomes. The assumed but hypothetical 'hidden orifice' then seems to have been regarded[3] as evidence for taxonomic relationship between the two orders.

Quite apart from the long time gap between the Permo-Triassic extinction of the cryptostomes and the appearance of the cheilostomes in the Cretaceous, there is a fundamental difference in the organization of zooids and colonies in the two orders. The cheilo-

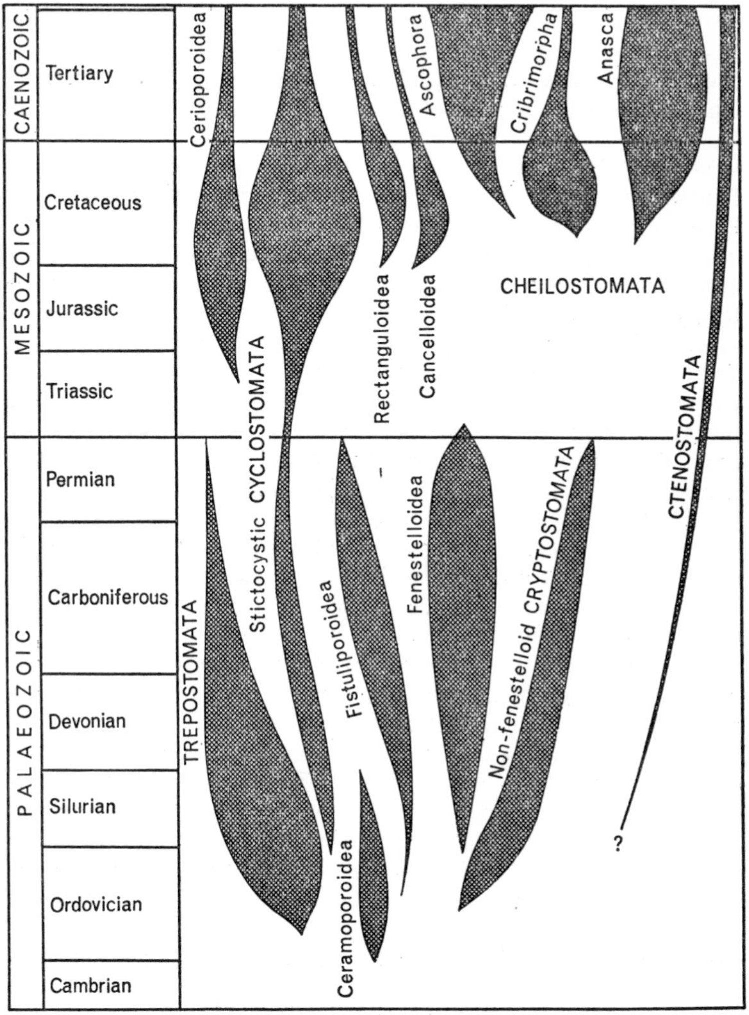

Fig. 20 Geological history of the Bryozoa.

stome colony increases by the growth of peripheral buds which remain, via the interzooidal pores, part of an internal continuum. This basic character of the cheilostomes is immensely important and has, *inter alia*, made possible the evolution, diversification and proliferation of non-feeding heterozooids. In cryptostomes, interzooidal communication was probably achieved through the hypostegal coelom of an external mantle, which is a structure quite lacking in the colonies of primitive anascan cheilostomes. The interzooidal crosswalls are so dissimilar in the two orders, that derivation of the cheilostome end-septum from the fenestelloid end-wall must be regarded as highly improbable.

Cyclostomata

Different by far, from that of the orders so far considered, was the Palaeozoic history of the Cyclostomata. Last of the stenolaemate orders to appear, they never enjoyed spectacular success, but maintained themselves in small numbers throughout the era. One superfamily, the Hederelloidea, was confined to the Palaeozoic; but two others, Tubuliporoidea and Articuloidea, survived the Permo-Triassic transition and carried the order into the Jurassic period. The only calcified bryozoa present in Middle Mesozoic seas, they started to diversify during the Jurassic (if not before);[136A] by the Lower Cretaceous their increase had become explosive, and their zenith during this period constitutes one of the highlights in the history of bryozoa. Their numbers dwindled during the late Cretaceous and early Caenozoic, presumably under impact of competition from the newly arisen and actively radiating cheilostomes, and in Recent seas they are numerically unimportant. As surviving stenolaemates, however, they hold great scientific value.

The superfamily Articuloidea has at no time contained many genera, and much of the Mesozoic success of the Cyclostomata is attributable to evolution within the Tubuliporoidea. The latter were represented during the Cretaceous by over 70 genera,[3] less than half of which survived into and through the Tertiary. Mesozoic diversity was boosted by the appearance of two more superfamilies, Salpingoidea and Dactylethroidea, both of which became extinct before the close of the era. The Salpingoidea appear to be a distinctive offshoot from the Tubuliporoidea, and will be considered in detail presently (p. 137); the Dactylethroidea, comprising four exclusively Cretaceous genera, are of more obscure affinities and require fresh interpretative studies.

Three other stenolaematous superfamilies arose during the Mesozoic, all—in contradistinction to the tubuliporoid line—characterized by coelocystic walls: Cerioporoidea, Cancelloidea and Rectangu-loidea, all of which deserve separate discussion.

Cerioporoidea

The cerioporoids appeared in the late Triassic. Their fortunes paralleled closely those of the continuing cyclostome lines (Tubuliporoidea and Articuloidea), and there are today few surviving genera. *Heteropora* is one of these. The Cerioporoidea, as previously explained (p. 110), differ from typical cyclostomes and resemble trepostomes in their ability to produce new zooids all over the colony surface. Either, then, the Trepostomata did not become extinct in the Permo-Triassic but tenuously survived to initiate another successful stock; or there evolved from the stictocystic Cyclostomata a new coelocystic line which independently developed the power to produce interzooidal buds. If the latter be true, then evolution repeated itself and virtually reconstituted the Trepostomata in Mesozoic times!

The proposal that the Cerioporoidea should be classified with the trepostomes is not new, but was given more force by Borg's[120] anatomical studies on *Heteropora*, which revealed the extent of the differences between the Heteroporidae and typical cyclostomes. There is, however, one feature in which the Cerioporoidea and Trepostomata do not agree.[5] Examination of the microstructure of trepostome walls suggests that the laminae from which they are constructed were deposited parallel to the secretory epithelium. So, in effect, the surfaces of laminae are growth surfaces, and the laminae themselves are interpreted in sections as growth lines. At the growing edge of an interzooecial wall, the laminae converge and overarch (Fig. 18B,D). Thin sections of heteroporoids, however,[5] do not display laminal convergence orally, and the laminae diverge as they approach the growing edge. Whatever the interpretation and significance of this arrangement, it certainly represents a notable difference between trepostomes and cerioporoids.

Earlier investigations of fossil cerioporoids have been directed mainly towards the compilation of catalogues of Jurassic and Cretaceous species.[129] These were, no doubt, a necessary preliminary; but studies on astogeny and wall microstructure, for example, have been neglected. Further work on the Mesozoic–Caenozoic Stenolaemata may well clarify the taxonomic position of the Cerioporoidea, including the Recent Heteroporidae, which meanwhile remains an open question.

Cancelloidea

The cancelloids appeared in the Jurassic and were especially successful during the Cretaceous: among stenolaemates only the Tubuliporoidea were then represented by a greater number of genera. Hardly any cancelloid genera survived the Mesozoic, however, and the superfamily is sparsely represented in Recent seas.

Colonies are chiefly ramose with more or less cylindrical branches. As explained in the previous chapter (p. 105), the external walls are coelocystic. The cancelloids are particularly characterized by the thick deposit of laminated, secondary calcification which, secreted by the hypostegal epithelium, covers much of the colony. The so-called cancelli (*cancelli* (pl.), lattice), from which the superfamily was named, have usually been interpreted as tubules[129] which, in species of *Hornera*, are claimed by Borg[119] to unite the exosaccal and hypostegal coeloms. However, Tavener-Smith[135] reports only solid rods traversing the secondary calcification in *Hornera frondiculata*. Whether or not tubular pseudopores are actually present—and there is no *a priori* reason why they should not be—the structure and mode of calcification in the Recent Horneridae is strikingly similar to that found by Tavener-Smith in fenestelloids (p. 126), which they evidently parallel in this respect.

Rectanguloidea

The superfamilial name refers to the orientation of the zooecia, the walls of which, instead of contributing to the frontal surface of the colony as in tubuliporoids, lie at right angles to it (at least to some extent). Recent representatives, in the family Lichenoporidae, were studied by Borg[119] and have been discussed in Chapter 6 (p. 109). The superfamily first appeared in the Lower Cretaceous, diversified little, and has few surviving genera.

Borg, in his last paper,[122] drew attention to the remarkably close morphological and astogenetic parallels between rectanguloids and the Palaeozoic fistuliporoids (p. 121). It is not quite clear whether or not Borg believed that the rectanguloids actually represent a Mesozoic recrudescence of the fistuliporoids. The long time interval (90 m y) between the assumed extinction of the latter in the late Permian and the first recorded appearance of the rectanguloids in the early Cretaceous makes this seem a little unlikely. The alternative solution, however, is that the rectanguloids represent another example of evolutionary repetition, which is hardly less remarkable. It is certainly curious that the Cerioporoidea, Cancelloidea and Rectanguloidea are all apparently characterized by methods of colony formation previously developed by Palaeozoic groups, all of which had meanwhile become extinct. The relationships of the three superfamilies with the stictocystic cyclostomes are far from clear, and it is to be hoped that further research in the relatively poorly known Lower Mesozoic will provide at least some of the answers.

Salpingoidea

Another interesting group of Mesozoic Stenolaemata is the Salpingoidea, with a known range from the Middle Jurassic to the Late Cretaceous. The possession by these forms of operculate zooecia and avicularium-like heterozooecia separate them sharply from other cyclostomes, and have led some workers to regard them as cheilostomes or their progenitors.

The careful study by Levinsen[131] on the genus *Meliceritites* is of great value in understanding the Salpingoidea. Colonies are incrusting and disciform, or erect with cylindrical branches or foliaceous fronds. The exposed walls are porous, having the usual appearance of a stictocyst. The zooecia are long and tubular, but differ from the usual cyclostome type in having the distal region abruptly and extensively dilated.

The colony surface is divided into more or less hexagonal areas, each corresponding to the exposed part of a zooecium and bearing an orifice, the superficial effect being remarkably cheilostome-like. The orifice may be shortly tubular or in the plane of the frontal wall, and has the shape of a deep D with the proximal margin straight. It is closed by a calcareous operculum marked by fine striae radiating from the centre of the proximal edge.

The brood chambers are typical cyclostome gonozooecia. The avicularium-like heterozooecia are basically like the autozooecia, but frequently have the orifice greatly attenuated in the distal direction. The mandible then has the form of a slender isosceles triangle, just as in cheilostomes.

The first of several questions prompted by a study of the Salpingoidea is: how did the operculum open? There was no frontal membrane and apparently no other form of compensatory hydrostatic apparatus; the system, therefore, must have been internally balanced, as in living cyclostomes. The operculum might simply have been pushed open by the emerging tentacles, at least in the first species to evolve.

The second question is: if opercula and avicularia are present, how can we be sure that the salpingoids are neither cheilostomes nor ancestral to them? First, the tubular zooecia, stictocystic walls and large gonozooecia all characterize cyclostomes, not cheilostomes. Second, neither the operculum nor the avicularium in the two groups are exactly comparable. The salpingoid operculum is a calcified lamina quite different from the structure encountered in cheilostomes; while the heterozooids lack any area covered by a frontal membrane, which is an essential feature of the cheilostome avicularium. If the apparent similarities are not based on homology, there are no valid grounds for regarding salpingoids either as cheilostomes or as their progenitors.

This conclusion itself poses further questions: what is the function of the avicularian type of heterozooid?—and why should so specialized a structure have evolved twice during bryozoan evolution? Unfortunately, we do not know. One possible function for large cheilostome avicularia, mentioned by Kaufmann,[1] which might also apply to the analogous heterozooids of salpingoids, was suggested by the observation that, in certain cheilostomes with an extended bathymetric range, both number and size of avicularia increase in proportion to the depth of the water. Thus, it was asked, might not a to and fro movement of the mandibles create currents in still water? Yet if the prime role of the avicularian mandible was as an oscillatory device, rather than as one for snapping shut, would not adductor and abductor muscles have approximately equal mass as in vibracula? In cheilostomes, at least, the adductors are very much larger. Other possible functions of avicularia have already been considered in Chapter 5 (p. 94), and none seems to offer a convincing explanation of why such heterozooids should have evolved once, still less twice!

GYMNOLAEMATA

Ctenostomata

The order Ctenostomata is reputedly of great age, with a fossil history stretching from the Lower Ordovician to the present time.[3] The records are based on shallow excavations, supposed to represent the position of stolons, discovered in brachiopod shells. In the Lower Ordovician fossils this interpretation appears open to question. A few of the finds, however, such as *Rhopalonaria*, which appeared first in the Upper Ordovician, bear a good resemblance to the impressions attributed to ctenostomes in later periods. Several of these fossils, such as *Ascodictyon* and further species of *Rhopalonaria* from the Carboniferous and Permian, have been well described and illustrated by Condra and Elias.[104]

Shell borings, which have been attributed—in an unpublished study by Dr Anna Hastings—to the extant genus *Terebripora*, are known from a British Jurassic locality, and are more frequently encountered in the Tertiary. Stoloniferous ctenostomes of late Cretaceous age, placed in a special genus *Stolonicella*, have been described by Voigt[113] from well-preserved moulds formed as the colonies were overgrown by incrusting organisms—in one case a *Lichenopora*—which were subsequently fossilized.

It can be concluded from the sparse fossil record that the Ctenostomata were in existence during the Jurassic, and may probably be traced back as far as the Ordovician. They definitely preceded the Cheilostomata.

The classification of the Ctenostomata proposed by Silén[85] recognizes two suborders: STOLONIFERA and CARNOSA. In the former, autozooids are borne on kenozooidal stolons; in the latter, true stolons are absent, though often simulated by pseudostolonal outgrowths from the autozooid bases. All the earliest fossil ctenostomes appear to have been stolonate, although study of present-day tunnelling forms[90] has shown that remarkably similar borings may be the product of stoloniferans, such as *Spathipora*, *Terebripora* and others (discussed in Chapter 4, p. 79), or carnosans like *Immergentia*. It might be unwise, therefore, to conclude that all stolonate fossil ctenostomes necessarily belong to the Stolonifera as defined by Silén. Furthermore, an undoubted carnosan from the Cretaceous, referred to the extant genus *Arachnidium*,[115] immured on a belemnite 'bullet' by overgrowth of a serpulid tube, has been discovered recently in north Germany.

Much of Silén's argument concerning the origin and relationships of Ctenostomata and Cheilostomata (see below) stems from the premise that the former order is diphyletic. Can this assumption be justified?

The contention appears to receive support from the ontogenetic studies of Soule,[92] who discovered that the chronology of muscle development in the newly formed zooid usually conforms to one or other of two distinctive patterns. One sequence characterizes the Carnosa, another the Stolonifera, as follows:

	A. Carnosa	B. Stolonifera
1st	Transverse parietals	Longitudinal parietals
2nd	Retractor	Retractor
3rd	Longitudinal parietals	Transverse parietals

There are, however, differences between the systems of Silén and Soule, in that the pseudostolonate families Nolellidae and Immergentiidae (Paludicelloidea) display developmental sequence B, which places them in the Stolonifera. It may be thought significant that, of Silén's Carnosa, it is those with the best developed pseudostolons that display the ontogenetic pattern typical of truly stolonate forms. Finally, proving that the stolonate and pseudostolonate colony types are not immutably distinct is the nolellid genus *Bulbella*, undescribed when Silén's work was published, in which the autozooids may be pseudostolonate or borne on true stolons.[31]

It may be concluded that, irrespective of whether or not the chronology of muscle development should be regarded as the overriding taxonomic criterion, there is no obligation to assume that stolonate and non-stolonate ctenostomes represent separate phyletic lines.

Origin and early history of the Cheilostomata

It was for some time believed that cheilostomes had been found in Upper Jurassic rocks, but Voigt[114] has established that such records were based on errors of identification or stratigraphic attribution. The first acceptable finds are of electrine anascans (e.g. *Pyripora*, *Rhammatopora*) from the Albian horizon of the Lower Cretaceous.[19]

Origin of the Cheilostomata from either the Cryptostomata or the salpingoid Cyclostomata has already been discounted. The evidence of comparative anatomy, in any case, points to a very close relationship between the Cheilostomata and the Ctenostomata (see Chapters 2 and 5), and the latter were already in existence at the start of the Cretaceous. The most obvious difference between cheilostomes and ctenostomes is the lack of calcification in the zooid walls of the latter. In fact, the walls of *Amathia* and *Bowerbankia* contain about 20% mineral matter (dry weight), of which a substantial part is attributable to calcium and magnesium salts.[82]

The most satisfactory hypothesis at the moment is to regard the cheilostomes as derived from a ctenostomatous ancestor of paludicelloid type; and the discovery of such a form, *Arachnidium*, from the Barremian horizon (below the Albian) of the Lower Cretaceous is an exciting and significant event. Although such sparse evidence could well be misleading, the fossil record at the moment is thus consistent with such an idea. *Arachnidium* and *Pyripora*—both extant genera—have a similar type of colony consisting of adherent, uniserial, branching chains of pyriform zooids, the latter slender proximally. The zooids of *Pyripora*, however, are calcified, and have a frontal membrane. *Rhammatopora* resembles *Pyripora* but bears a series of spines around the frontal membrane; neither (cf. *Callopora*) has ovicells.[110]

Silén[85] suggested the derivation of both Ctenostomata and Cheilostomata from an ancestral type resembling *Labiostomella*. In this interesting and peculiar anascan genus the colony is erect, with branches composed of zooids aligned in single series. Each zooid is, in shape, rather like that of *Paludicella* (pp. 31, 86), but has a distinct frontal membrane and a two-lipped orifice. The branches arise vertically from the ancestrula as derivatives of frontal buds, which have a bilaterally symmetrical disposition on the ancestrula, exactly similar to that of the spines which surround the frontal membrane in *Callopora* or *Rhammatopora*. The precise significance of this, in phylogenetic terms, is difficult to assess, for there is no indication in the fossil record of primitive anascans with erect colonies. It is, moreover, difficult to relate the *Labiostomella* type of ancestrular budding to that found in any other extant gymnolaemates.

Once established, the Cheilostomata underwent explosive development and diversification. The inferred evolutionary progression indicated in Chapter 2, and based on comparative ontogeny and astogeny of living species, is supported by the sequential appearance of structural types in the Upper Cretaceous, as revealed especially by the studies made over a period of years by Professor Voigt of Hamburg.[112,113]

Thus the Malacostegoidea, from the late Lower Cretaceous (Albian), were the first anascans to appear, closely followed by the Coelostegoidea—notably the family Onychocellidae. The Cribrimorpha, which were sufficiently successful to justify separate discussion presently, also evolved early. The more advanced anascans appeared during the early Tertiary. Some of these, such as the Cellularioidea, are lightly calcified and may not, therefore, have left a representative fossil record; although some remarkably delicate cheilostomes have been recorded from the early and mid-Tertiary by Lagaaij.[1]

The Ascophoran family Hippothoidae, generally considered primitive and rather distinct, dates from the early Upper Cretaceous (Cenomanian). More typical ascophoran families, such as Escharellidae, Hippoporinidae and Schizoporellidae, had appeared by late in the period. The gymnocystidean family Exochellidae was first recorded in the middle of the Upper Cretaceous. No detailed analysis of the pattern of ascophoran and gymnocystidean evolution is possible until more work has been carried out on the cheilostome frontal wall, and the results applied to the classification of both fossil and Recent species.

Cribrimorpha

The abundance and accessibility of the English Chalk (Upper Cretaceous) formations has naturally facilitated study of the bryozoan fossils they contain. The colonies of cribrimorphs may be found incrusting pieces of echinoid test or shell fragments of the bivalve *Inoceramus*. Specimens can be cleaned by soaking in water and stroking with a fine brush, but brushing can damage delicate structures. A better method, with which some workers have obtained beautiful preparations, is to place the specimen below a dripping tap, allowing the force of the falling drops to free and wash away argillaceous matter. Immersion for some minutes in saturated salicylic acid, followed by prolonged washing, has been found effective for loosening adherent chalk matrix.[109] A modern technique, suitable only when the matrix is softer than the fossil, is to make use of an ultrasonic vibrator operating at 40,000 cycles/sec. The specimen is placed in a test tube partly filled with water, to which a few drops of Tepol have been added. The tube stands in the tank of the vibrator, clear of the bottom, and with the water surface inside the tube level with that outside. Soft, fine material is rapidly removed from the specimen and forms a suspension in the water. Temporary staining with indigo water-colour is another valuable technique,[108] for the colour penetrates the substance of the specimen and differentiates structures according to their texture.

Colonies are occasionally bilaminar, but more usually incrusting. The zooecia sometimes form branching, uniserial rows; but more frequently the colony is fan-shaped at first, becoming more or less circular as astogeny proceeds.

The distinctive features of cribrimorphs are zooecial (Chapter 2, p. 38). Much of the exposed surface is made up from the *frontal shield* of medially fused ribs or *costae*, which cover the frontal membrane (Fig.5C, p. 39); proximal and lateral to the membrane is a variously developed gymnocyst from the edge of which the costae arise. Adjacent costae may be joined by a series of lateral fusions, leaving rows of intercostal pores, which may be clearly linear or rather irregularly arranged.

Presumably, as in the Recent Cribrilinidae, each costa enclosed a slender lumen; while its frontal face may incorporate pseudopores or *pelmata*. Thus the perforate appearance of a cribrimorph frontal shield is due partly to intercostal pores and partly to intracostal pelmata. In Cretaceous species[108,109] the costae are mostly linear and the lateral fusions between them have the form of regularly repeated tubular bridges, as in the Pelmatoporidae.

The fused pair of costae immediately proximal to the orifice constitute the *apertural bar*, which is differentiated in various ways. It may be fused with non-costate oral spines, so covering the orifice as a Y-shaped *oral shield*. A further characteristic of some Cretaceous cribrimorphs is the development of a layer of secondary calcification spreading from the interzooecial sutures over the gymnocyst, around the orifice and even over the frontal shield. Neanic zooecia usually resemble ephebic zooecia, but are smaller.

The geological history of the Cribrimorpha is one of brief but spectacular success. The most primitive species appeared in the early Upper Cretaceous, at the base of the English Chalk, not much later than the first malacostegoideans. These early cribrimorphs gave rise to some half-dozen families which flourished during the middle of the Upper Cretaceous, and were replaced by a smaller number of families during the late Cretaceous. All but two families (Cribrilinidae and Pelmatoporidae, which have extant members) had become extinct by the Eocene.[19] The number of genera has declined steadily through the Caenozoic.

Caenozoic Cheilostomata

Cheilostomes dominated the bryozoan fauna through the Tertiary, as they do today. The Anasca, excepting the cribrimorphs, maintained their position of importance; while the Ascophora and Gymnocystidea continued to diversify, thereby achieving the success they enjoy at present. Most ascophoran families had evolved by the close of the Eocene, and the specific composition of genera found in fossil assemblages becomes steadily more similar to that of Recent seas.

Tertiary cheilostomes are perhaps the most studied of all bryozoa, and the subject of many fine monographs. One of the first of these was George Busk's account of the bryozoa from the Coralline Crag (Pliocene) of East Anglia.[99] A representative selection of more recent important works has been included in the bibliography.[96–103,105,107,111]

8

PHYLACTOLAEMATA

BRYOZOA IN FRESH WATER

Freshwater bryozoans belonging to the class Phylactolaemata, though lacking in numbers and diversity, are widely distributed ecologically and geographically. They differ from the marine gymnolaemates and stenolaemates not only by virtue of their special adaptations to life in fresh water, but also in their basic anatomy. Many of their structural characteristics have been interpreted by comparative anatomists as primitive. Both Brien,[7] who has carried out important studies on phylactolaemates, and Hyman[17] have written comprehensive accounts of the class in their well-known textbooks, while those who read German have access to two more.[10,144]

Structure of the zooid

Phylactolaemates have not evolved polymorphism, so that all zooids are autozooids. Despite their several class characteristics, the zooids conform recognizably to the general bryozoan type with lophophore, deeply looped alimentary canal, conspicuous retractor muscles and dorsal ganglion. One immediately obvious difference, in expanded zooids, lies in the shape of the lophophore. In gymnolaemates the lophophore is circular, with the mouth in its centre; in phylactolaemates (other than *Fredericella*, in which it is almost circular), on the other hand, it is horseshoe-shaped, with the tentacles arranged along its margins so that it comprises a double series of tentacles arranged around a deep crescent. The mouth is situated in the bend of this U, the convexity ventral to it and the concavity dorsal to it, so that the two arms project on the dorsal side. These extensions make the lophophore larger than it is in other bryozoans, and it arries more tentacles—sometimes 100 or more. Those forming the nner series in the horseshoe are shorter than those making up the uter series. In each arm there is a ciliated groove between the rows f tentacles: the two grooves converge towards the mouth. The bases f the tentacles are connected by a thin intertentacular membrane.

The lophophore is regarded as constituting the *mesosome* of the ody, and its cavity is the *mesocoel.* A septum between the pharynx nd the ventral body wall partially divides the mesocoel from the nain body cavity or *metacoel* (Fig.21A). Projecting over the mouth rom the dorsal side of the lophophore is a hollow flap of tissue, the *pistome.* The epistome has coelomic continuity with the lophophore, ut is separated from it by short mesenteries which may perhaps be egarded as vestiges of a septum. The epistome is believed to represent he *protosome* (of *Phoronis*, for example) and its cavity the *protocoel.* This organ thus constitutes a primitive structure lost in other ryozoans.

The mesocoel extends into the arms of the lophophore, enters each entacle as its lumen and loops around the ventral side of the harynx. As it encircles the pharynx dorsally it encloses the ganglion nd merges with the protocoel. As union with the three mediodorsal entacles in the concavity of the lophophore is obstructed by the pistome, two extensions of the mesocoel (the *forked canal*) skirt the pistome, join—dilating to form a small sac—and divide, passing nto the tentacles.

The mouth leads to the U-formed alimentary canal which is lifferentiated into the usual parts: the pylorus, however, is not iliated. The caecum is connected to the ventral body wall by the uniculus, which is hollow and incorporates muscle fibres. There are o subsidiary funiculi.

The body wall is histologically well developed. The epidermis is omposed of columnar cells overlain either by a hard cuticle or by a hicker gelatinous layer. The epidermis is underlain by a layer of mooth circular muscle fibres, a basement membrane, a layer of mooth longitudinal muscle fibres and, finally, a peritoneum. Muscular layers, as opposed to bands, are absent from the walls of oth classes of marine bryozoa.) The body wall invaginates distally o form an atrium and the introvert (tentacle sheath).

Apart from the muscles of the body wall and the retractors, the rincipal muscles in the zooid are the serially arranged dilators of he atrium and the sphincter. Since the body wall contains muscular ayers, parietal muscles are correspondingly lacking. Eversion of the olypide is thus brought about by contraction of the circular muscles n the wall. Retraction is achieved in the usual way.

Nerve tracts from the ganglion encircle the pharynx and pass into the two arms of the lophophore, branching into the tentacles. The three mediodorsal tentacles (those linked to the forked canal) are innervated from a separate epistomial ring. There are no special sense organs.

The ovary is located on the ventral body wall in the distal part of the zooid; the testis or testes are usually borne on the funiculus.

The zooids are obviously cylindrical in *Fredericella* and separated from each other by septa. In *Plumatella* the septa are usually lacking, so that the coeloms freely communicate. In both of these genera the colonial budding pattern produces ramifying lines of zooids. In *Lophopus*, *Lophopodella* and *Pectinatella* the colonies form more compact gelatinous masses, in the latter reaching 0·5 m across. The polypides are here suspended in a single extensive coelom, and the septa are either incomplete or missing altogether. In *Cristatella* the slug-like colonies, several centimetres in length, are not permanently affixed to the substratum, but can creep about on the flat under-surface.

Sexual reproduction and embryology

Sex cells ripen during the summer. Spermatozoa develop around cytophores, as in gymnolaemates (the cyclostomes, therefore, stand apart in this respect). The mature sperm, however, differ from those of gymnolaemates (including the freshwater *Paludicella*) in the structure of all three components. The head is short, the middle region exhibits a spiral structure arising presumably from the arrangement of mitochondrial material, and the tail contains more cytoplasm. Such spermatozoa diverge considerably from the primitive type.[13] Fertilization has not been witnessed.

After ovulation the egg is transferred to an internal embryo sac where it develops. It is not known how transference is accomplished. The embryo sac arises, just above the ovary, as an invagination of the

---

Fig. 21 Structure in the Phylactolaemata

A Diagrammatic longitudinal section through a generalized phylactolaemate zooid. The development of new polypides from buds *a*, *B* and *C*, and of statoblasts (s1 to s8) is explained in the text

B Sessile statoblast of *Fredericella*

C Floating statoblast of *Lophopus* (side view)

D Spiny statoblast of *Cristatella*

a adventitious bud; *B* main bud; *C* duplicate bud; *fun b* funiculus of main bud; s1–s8 developmental series of statoblasts, all in sagittal section except s7 which is in horizontal section.

(A based on Brien[7] and Cori;[10] B–D after Allman[137]).

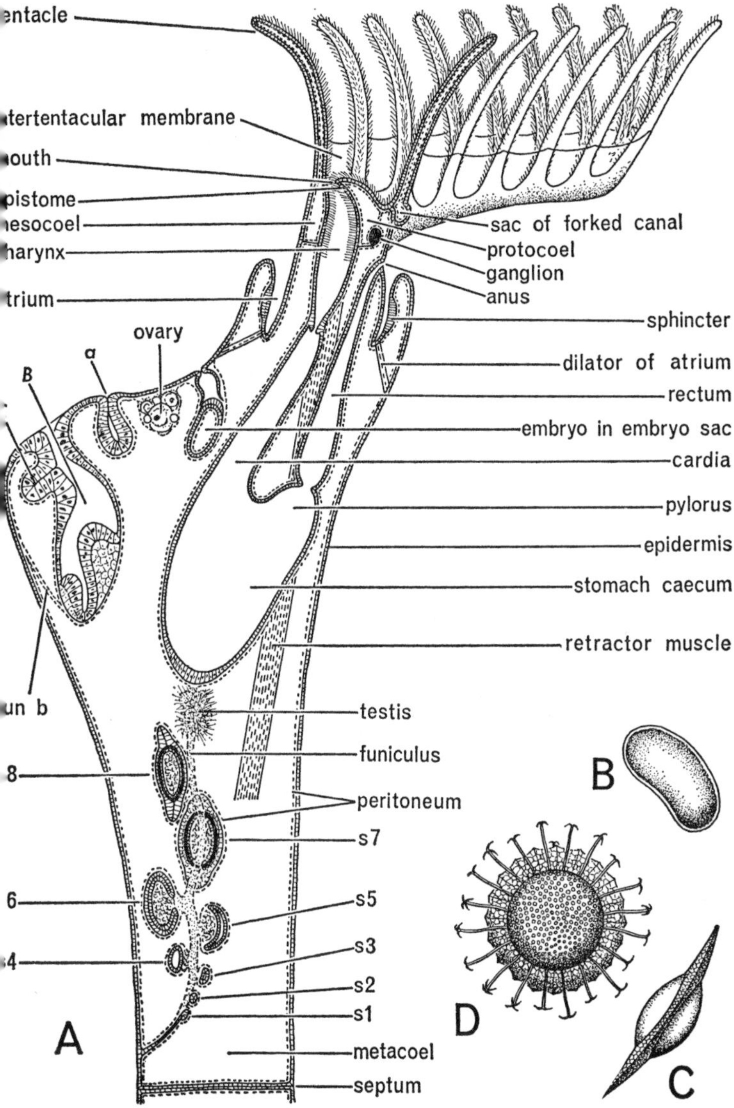

entacle
tertentacular membrane
outh
pistome
esocoel
harynx
trium
ovary
a
B
un b
8
6
4
sac of forked canal
protocoel
ganglion
anus
sphincter
dilator of atrium
rectum
embryo in embryo sac
cardia
pylorus
epidermis
stomach caecum
retractor muscle
testis
funiculus
peritoneum
s7
s5
s3
s2
s1
metacoel
septum
A
B
D
C

Fig. 21

body wall in which both cell layers participate. Its development (as in the ooecium of cheilostomes) appears to be a response to an ovarian hormone; unless it receives an egg, the embryo sac degenerates.

Cleavage proceeds as in gymnolaemates. How the hollow blastula gastrulates is still not clear, although earlier workers believed that there was an inwards movement of cells at the vegetal pole. Whatever the process, the ectoderm becomes lined by a peritoneum and the embryo becomes, in essence, an undifferentiated cystid. There is no entoderm. Still within the embryo sac, to which it may be attached by special cells, the cystid produces its first polypide—sometimes more than one—by invagination in the usual bryozoan manner. Meanwhile, below the blastogenic area, an annular *mantle fold* of the body wall is produced. The mantle fold advances distally, enclosing the polypidial region completely except for a small terminal opening and develops cilia. The embryo in this way becomes a 'larva' which is, in effect, a juvenile colony. Larval ontogeny is clearly so specialized that it provides no phylogenetic clues.

The larva escapes through an opening in the external wall of the embryo sac, and is active for a short period. As it swims, the pole opposite to the polypides leads. At this pole an aggregation of sensory cells and nervous tissue has been reported, possibly corresponding to the apical sense organ in gymnolaemate larvae. The larva, which has a negative response to light, locates and affixes to some suitable substratum by its leading pole; the mantle rolls back and the polypides emerge through its opening.

Budding and astogeny

The newly attached juvenile colony expands by producing more polypides. When, as in *Plumatella*, the larva contains two polypides colony growth proceeds from the outset in two diametrically opposed directions. A twin or '*jugalis*' (*iugalis*, yoked together) type of colony results (quite different from the situation in the cheilostome genus *Membranipora*, which has a twin ancestrula which buds in one direction only). In *Fredericella* and *Plumatella*, which have a branching type of colony, polypides arise in ordered manner from a blastogenic zone on the ventral side of the maternal polypide represented by the letter *A*. Three rudiments are present simultaneously: the *main rudiment B*, a *duplicate rudiment C* arising from the ventral side of the main one, and, nearer to the maternal polypide an adventitious rudiment *a* (Fig.21A).

As the main rudiment *B* becomes a fully formed polypide, so its duplicate *C* separates and becomes the next main rudiment in line at once producing its own duplicate *D*), while a new adventitious rudiment arises between *B* and *C*; and so on. Meanwhile the adventitious rudiment *a* has become a main rudiment *B′* (at once producing its own duplicate *C′*) while another adventitious rudiment arises between *A* and *B′*; this succession from *A* also continues, though only for a limited number of generations before the original polypide dies.

The sequential production of polypides just described causes the colony to bifurcate at every zooid, so that a much divided system of uniserial branches results.

In genera with more compact colonies budding proceeds in basically the same way, but there are differences in detail. In *Lophopus* the main rudiment appears to the left of the median sagittal plane of the maternal zooid, the adventitious rudiment to the right of it. The colony thus becomes fan-shaped.

Production of a new cystid, in so far as cystids are present, follows inception of the polypide, as in living stenolaemates but contrary to the sequence found in gymnolaemates.

During polypide inception, according to Brien,[7,138] the epidermal cells dedifferentiate and proliferate, giving rise to a locus of blastogenic tissue. This invaginates, carrying the peritoneum with it (*a* and *C* in Fig.21A). The epidermal lining of the rudiment develops into the gut epithelium (*C* in Fig.21A), from which the ganglion arises as an invagination of the dorsal wall of the future pharynx. The lophophore and epistome develop as coelomic outpushings of the cell layers. The funiculus (fun b in Fig.21A) arises from the peritoneum as the invaginating rudiment separates from the zooid wall. It is at first a continuous strand, but becomes hollow; some of the blastogenic epidermal cells migrate into its lumen.

Statoblasts

Earlier (Chapter 4, p. 71), the resting zooids or *hibernacula* produced by gymnolaemates living in fresh waters were described. Phylactolaemates also produce dormant bodies, but these are entirely different from hibernacula and are termed *statoblasts* (*statos*, standing, i.e. quiescent, +*blastos*, bud).[137] A statoblast is an asexually produced bud which arises from the funiculus, especially towards autumn, and acquires a hard protective case.

A statoblast is initiated—and in *Plumatella* a series of them develops at one time—when some of the blastogenic cells which invaded the funiculus clump together as a ball (s1 in Fig.21A), solid at first (s2) but hollow later (s3, s4). The spheres of cells increase in size, swelling out from the funiculus like cysts, while remaining covered by a layer of cells from the funicular peritoneum. The sphere of epidermal cells becomes compressed, giving rise to a hollow disc at the pole furthest from the funiculus (s5); meanwhile peritoneal cells invade the developing statoblast, occupying the space available near the funiculus. These cells become filled with yolk spherules. The epidermal cells, now transformed into a two-layered plate, overgrow and enclose the yolk-filled peritoneal cells (s6). There is now a double epidermis around the yolk cells. The inner epidermis—destined to become the epidermis of the future zooid—retains the appearance it has at this stage; but the cells of the outer epidermis grow columnar and start secreting a shell from their inner surface. The germinal mass thus becomes encapsulated by a chitinous layer situated between the inner epidermis and the outer epidermis (s7).

The protective covering of the statoblast becomes further differentiated, and acquires a form that varies from genus to genus. First, the shell develops an equatorial suture, so that dorsal (the hemisphere formerly occupied by the epidermal disc) and ventral (the hemisphere which became filled by yolk cells) valves may be distinguished. In those statoblasts which float on release, found in all genera except *Fredericella*, the cells of the outer epidermis bordering the suture elongate greatly. During this process the cell contents disappear and the cell walls harden: the result is a crest or *pneumatic annulus* of air-filled cells, which encircles the *protective capsule* containing the germinal mass (s7). Non-floating statoblasts may also be produced, in which the annulus is present but is not pneumatic.

The statoblasts of *Fredericella* (Fig.21B) lack the annulus and do not float. They may develop attached to the body wall (and be stuck, with it, to the substratum) or free in the body cavity. The former (*fixed statoblasts*) remain *in situ* and thus serve to maintain the species in the occupied location; the latter (*free statoblasts*), which lack any adhesive structure, serve for dispersal. In other genera the statoblasts are *annulate*, and may be *floating* or *fixed* depending upon whether the annulus contains air or not; in the second case the statoblast is adhesive. In *Lophopus* (Fig.21C) and *Plumatella* the annulus is unadorned, but in several genera including *Cristatella* (Fig.21D) and *Pectinatella* the annulus bears hooked spines. The ventral valve is generally more convex than the dorsal valve, and the annulus slopes dorsally towards its periphery (Fig. 21A,s8;C), so that the dorsal side floats uppermost in the water.

Non-floating statoblasts have been named *sessoblasts* if cemented to the maternal tissues and *piptoblasts* if they are free; floating statoblasts without spines have been called *floatoblasts*, those with spines *spinoblasts*. The terms provide a useful shorthand but are ugly, and, as Lacourt points out in his recent *Monograph of the freshwater Bryozoa-Phylactolaemata*,[143] have the disadvantage of eliminating the descriptive prefix *stato*. Moreover, it is by no means certain that the grouping implied by the four names is really the natural one.

In *Cristatella*, *Fredericella*, *Lophopus* and *Pectinatella* there is usually only one statoblast formed by each polypide; in *Plumatella*, on the contrary, a number form in orderly succession down the funiculus proceeding away from the stomach.

The number of statoblasts produced in favourable environments is remarkable. Brown[139] reported drifts of statoblasts up to and even exceeding 1 m in width extending along half a mile of shore of Douglas Lake, Michigan, and calculated that colonies of *Plumatella* in 1 m^2 of the plant zone might release 800,000 statoblasts in the autumn.

Statoblasts may be ejected from the parent zooid via an atrial pore during the summer. More often they accumulate through the growing season and are liberated when the colonies disintegrate at the approach of winter. Statoblasts released in summer can germinate after a short dormant period, but those liberated during the autumn usually overwinter before germinating. Thus, in temperate latitudes, there may be two generations arising from statoblasts each year. Prolonged desiccation usually destroys the viability of statoblasts although those of *Lophopodella* have germinated after being dry for 50 months. They are resistant to freezing. Frozen statoblasts may remain viable much longer, and ultimately germinate more rapidly, than those which have been dried and stored at room temperature.

The annual cycle in temperate countries appears to be largely under the control of temperature, with the colonies generally dying down at the onset of winter. The statoblasts survive, and germinate in spring as the temperature rises.

When germination commences, the two valves of the statoblast dehisce along the suture, while the germinal mass expands and gives rise to a single polypide in the usual way by invagination of a vesicle. As the yolk is utilized, the storage cells revert to normal and become the peritoneum of the first cystid. The first polypide emerges from between the valves, and others soon follow, so that a young colony is produced before the valves are lost. The colony adheres to the substratum by a secretion coming from the epidermis opposite the polypide rudiment.

Degeneration of polypides

Phylactolaemate polypides degenerate in much the same way as those of gymnolaemates and for similar reasons, but the polypide disintegrates more fully so that brown bodies do not form. No replacement polypide is produced in the same cystid, but new polypides are arising continuously in the growth zones of the colony.

Ecology

Although the freshwater bryozoa are by no means uncommon, they can seldom be found without a careful search. Colonies may be found during summer in almost every lake, pond, river and stream, especially in clear, quiet, shallow water containing plenty of plants. Their presence may often be detected from the occurrence of statoblasts floating at the top of the water, where they can be discovered by skimming the surface with a fine hand-net.

Still waters are generally the most favourable, although *Plumatella emarginata* (in contrast to *P. repens*) apparently prefers flowing water, at least in some localities.[141] *Fredericella sultana* has been found in lakes at considerable depths, and even in less deep lakes it generally occurs nearer the bottom than other phylactolaemates.[141] The colonies of other species are to be sought in shallow water, and are generally found adhering to tree roots or to the shaded side of twigs, floating branches or aquatic plants and, as in marine bryozoans, there is clear evidence of preferred substrata. The colonies may be found on the underside of the floating leaves of water lilies (*Nuphar*, *Nymphaea*) or pondweeds (*Potamogeton*)—where they are insulated from the adverse effects of a fluctuating water level—less often on the stems of reeds (*Scirpus*, *Typha*, etc.), on the moss *Fontinalis* or on algae. They occur rarely on stoneworts (*Chara*).

The colonies of *Cristatella*, unlike those which are permanently fixed, have the ability to move about. The colony consists of an elongated mass of greenish, gelatinous substance, which, when well grown, may reach or exceed a length of 20 cm, with a transverse diameter of about 1 cm. It crawls on a flattened sole, and the polypides project from its upper surface. A distance travelled of 10–15 mm per day seems usual, although larger colonies may not move at all. Limited movement may also occur in *Lophopus* and *Pectinatella*, generally as daughter colonies produced by lobulation draw apart. The locomotory mechanism has not yet satisfactorily been resolved, but in *Cristatella* the body-wall musculature is probably involved, for the muscles are especially well developed in the creeping sole.

Alkaline waters seem more suitable than acid ones, especially in the pH range 7–9. In the British Isles the Broads area of East Anglia is not only particularly favourable, but also the best studied.[147] Several species are here widely distributed, noticeably *Fredericella sultana*, *Lophopus crystallinus* and *Plumatella repens*. *Cristatella mucedo* favours clear, well-lit waters, and was formerly very plentiful. E. A. Ellis in his book *The Broads* reports a disturbing decline in the abundance of this species in recent years: it appears that pollution and the increasing turbidity of the broadland waterways during the summer months are inimical to it. Evidence from several sources suggests that species of *Plumatella* are the most resistant to pollution.

There is a surprising dearth of published information regarding the presence and distribution of freshwater bryozoans in the English Lake District. However, Dr Anna Hastings, in collections made some years ago, found that *Fredericella sultana*, *Plumatella fruticosa* and the gymnolaemate *Paludicella articulata* were widespread in the lakes and tarns. A second species of *Plumatella* occurred only in the physiographically and ecologically most developed lakes, while *Cristatella mucedo* was found only in the Windermere drainage system. Colonies were located on logs and dead branches, but more especially on the under surface of stones. Suitable stones were always resting on other stones, for the bryozoans require to be clear of mud and silt. In a recent American study, Bushnell[1] has reported seven species from montane lakes in Colorado.

The former presence of bryozoans in the water mains of Manchester and elsewhere also testifies to the widespread distribution of many species. Harmer,[142] who summarized the occurrences of bryozoans in water works, noted that *Fredericella* and *Plumatella* in particular had been recorded from the pipes of many water supply systems—often in staggering quantities (see p. 12)—prior to the adoption of sand-filtration as standard practice. The efficiency of this technique depends on the biological film which comes to coat the sand surface. Not only are such relatively large bodies as statoblasts removed, but also the microscopic organisms on which the filter-feeding bryozoans depend for food. The use of sand filters for water purification was first advocated quite early in the nineteenth century, but the piping of unfiltered water was still causing blockages—due mainly to bryozoans—at the start of the twentieth century.

The mechanism of ciliary feeding is comparable to that of gymnolaemates, except that the moving water impinges on the groove which is present in each arm of the lophophore; cilia lining the groove convey particles to the mouth.[9] The freshwater bryozoans capture a wide variety of food organisms: bacteria, desmids, diatoms, flagellates, rotifers and minute crustaceans.

The species of phylactolaematous bryozoa tend to be distributed widely,[140,143] and many are cosmopolitan (presumably on account of the ease with which statoblasts can be disseminated). The ecology of one of these species, *Plumatella repens*, has been studied in detail in two lakes in Michigan.[141]

By marking individual colonies and zooids in the field, it was established that there is much variation in the longevity of individual polypides: 20% survived for four weeks, almost 50% for three weeks. In a graph of survival ($\log_e$ total number of surviving polypides against age in days) a theoretical maximum of about 53 days was indicated. Water temperature at this time was in the range 14–31°C.

The same methods provided proof that polypide formation in young colonies of *P. repens* followed a geometric rate of increase up to an age of six weeks (some 900 polypides): the relationship between $\log_e$ number of polypides and time was perfectly linear. The number of polypides N at time t is thus given by

$$N_t = N_o e^{rt}$$

where N_o is the initial number of polypides, e is the base of natural logarithms and r is the instantaneous rate of growth. Colonies doubled in size every 3½ to 5 days, or somewhat longer in spring when the temperature was lower. There was evidence of a decline in growth rate towards the end of the observational period, and that this might be correlated with the onset of sexual reproduction and statoblast formation (compare p. 85).

It is unfortunate that, despite the early example of G. J. Allman's *Monograph of the fresh-water Polyzoa*,[137] studies on British phylactolaemates have been rather neglected in recent years. This is all the more regrettable in view of the wide range of freshwater habitats here available within a relatively small area, and the ever increasing impact of an industrial civilization on our countryside and the natural environment. With a new identification manual available,[143] worthwhile ecological projects could be readily undertaken.

Phylogenetic considerations

The phylactolaemates are generally regarded as the most primitive of the living bryozoa. Like stenolaemates (and some ctenostomes) they preserve the cylindrical shape of the zooid, but phylactolaemates alone have a muscular wall with which to cause eversion of the lophophore. The latter is horseshoe-shaped in adult zooids, whereas in other bryozoans bilateral symmetry is seen only during ontogeny. Phylactolaemate zooids alone display the three body regions, protosome, mesosome and metasome. Their zooids are monomorphic. Phylactolaemates and living cyclostomes have in common a budding process in which the initiation of a polypide rudiment precedes formation of a cystid. Spermatozoa are more modified than those of cyclostomes, however; and the embryology—as in many freshwater representatives of predominantly marine groups—has been greatly modified.

Evolutionary trends within the class are fairly clear. Simplest is *Fredericella* with cylindrical zooids, covered by a cuticle, arranged in branching lines and separated from each other by septa. Only non-annulate statoblasts are produced. Morphological trends lead through a branching colony form with usually non-septate, sometimes gelatinous zooids, as in *Plumatella*, to the massive gelatinous colonies of *Pectinatella*, and may be said to culminate with the motile, slug-like *Cristatella*. Tentacle number increases through this series, from 20–30 in *Fredericella* to about 100 in *Pectinatella* and *Cristatella*. Similarly, non-annulate statoblasts are replaced by fixed or floating annulate statoblasts, and finally by spiny statoblasts (*Pectinatella* and *Cristatella*).

APPENDIX:

A CLASSIFICATION OF THE PHYLUM BRYOZOA

(including all genera mentioned in this book)

Although possibly unfamiliar to bryozoologists, the superfamilial suffix -oidea has been used throughout this book, as in the *Traité de Zoologie*,[7] in accordance with Recommendation 29A of the *International Code of Zoological Nomenclature* (1961).

† extinct taxa

Class **PHYLACTOLAEMATA**

Cristatella, Fredericella, Lophopodella, Lophopus, Pectinatella, Plumatella

Class **STENOLAEMATA**

Order *CYSTOPORATA*†
Superfamily Ceramoporoidea
(?) *Archaeotrypa, Ceramopora, Ceramoporella*
Superfamily Fistuliporoidea
Cyclopora, Fistulipora

Order *CYCLOSTOMATA*
Superfamily Tubuliporoidea
Berenicea, Diplosolen, Entalophora, Frondipora, Stomatopora, Tubulipora
Superfamily Hederelloidea
Superfamily Articuloidea
Crisia

Superfamily Cancelloidea
Hornera
Superfamily Cerioporoidea
Heteropora
Superfamily Dactylethroidea†
Superfamily Salpingoidea†
Meliceritites
Superfamily Rectanguloidea
Disporella, Lichenopora

Order *TREPOSTOMATA*†
Constellaria, Hallopora, Leioclema, Prasopora

Order *CRYPTOSTOMATA*†
Superfamily Rhabdomesoidea
Rhombopora
Superfamily Ptilodictyoidea
Sulcoretepora
Superfamily Fenestelloidea
Archaeofenestella, Archimedes, Fenestella, Fenestrapora, Polypora

Class **GYMNOLAEMATA**

Order *CTENOSTOMATA*
Suborder STOLONIFERA
Superfamily Walkerioidea
Hypophorella, Triticella, Walkeria
Superfamily Vesicularioidea
Amathia, Bowerbankia, Penetrantia, Spathipora, (?) *Stolonicella,*† *Terebripora, Zoobotryon*
Stolonifera incertae sedis: *Ascodictyon,*† *Rhopalonaria*†
Suborder CARNOSA
Superfamily Paludicelloidea
Arachnidium, Bulbella, Immergentia, Monobryozoon, Nolella, Paludicella, Victorella
Superfamily Halcyonelloidea
Alcyonidium, Flustrellidra

Order *CHEILOSTOMATA*
Suborder ANASCA
Superfamily Inovicelloidea
Aetea

Superfamily Scruparioidea

Eucratea, Labiostomella, Scruparia

Superfamily Malacostegoidea

Aspidelectra, Callopora, Carbasea, Cauloramphus, Chartella, Conopeum, Crassimarginatella, Electra, Flustra, Membranipora, Membraniporella, Pyripora, Ramphonotus, Rhammatopora,† Sarsiflustra, Tendra

Superfamily Cellularioidea

Beania, Bicellariella, Bugula, Caberea, Himantozoum, Kinetoskias, Levinsenella, Notoplites, Scrupocellaria

Superfamily Coelostegoidea

Calpensia, Cupuladria, Discoporella, Floridina, Lunulites,† Micropora, Steginoporella, Thalamoporella

Superfamily Pseudostegoidea

Cellaria

Suborder CRIBRIMORPHA

Cribrilina

Suborder GYMNOCYSTIDEA

Escharoides, Metrarabdotos, Umbonula

Suborder ASCOPHORA

Adeonella, Celleporina, Crepidacantha, Cryptosula, Escharella, Escharina, Fenestrulina, Haplopoma, Hippopodinella, Hippoporidra, Hippothoa, Iodictyum, Margaretta, Microporella, Myriapora, Pentapora, Porella, Savignyella, Schizobrachiella, Schizomavella, Schizoporella, Sertella, Smittina, Smittoidea, Watersipora

REFERENCES

The list includes a selection of important works which are of value to those interested in the Bryozoa, with an emphasis on recent publications. All are in English unless otherwise indicated (F—French, G—German). The International Bryozoology Association's 1st Conference Proceedings[1] contains many papers not otherwise separately listed, although—to facilitate reference to this volume—authors' names have been mentioned in the text whenever appropriate. An asterisk denotes a work with an extensive bibliography.

GENERAL WORKS

1. Annoscia, E., Ed., 1969. *Atti Soc. ital. Sci. nat.*, **108**: 1–377. (Proceedings of first conference on Bryozoa)
2. Barrois, J., 1877. *Recherches sur l'embryologie des Bryozoaires.* (F)
3. Bassler, R. S., 1953. *Treatise on invertebrate paleontology,* Ed. R. C. Moore, Part G. (Definitions of all bryozoan genera)
4. Becker, G., 1937. *Morph. Ökol. Tiere,* **33**: 72–127. (Digestion. G)
5. * Boardman, R. S., and Cheetham, A. H., 1969. *J. Paleont.,* **43**: 205–233. (Review of growth, variation and evolution)
6. Boardman, R. S., and Utgaard, J., 1964. *J. Paleont.,* **38**: 768–785. (Acetate replicas of polished sections)
7. * Brien, P., 1960. In *Traité de Zoologie,* Ed. P.-P. Grassé, **5** (2): 1054–1335. (F)
8. Brien, P., and Papyn, L., 1954. *Annls Soc r. zool. Belg.,* **85**: 59–87. (Comparison of Bryozoa and Entoprocta. F)
9. Bullivant, J. S., 1968. *N.Z. Jl mar. freshwat. Res.,* **2**: 135–146. (Feeding in lophophorates)
10. * Cori, C. J., 1941. In *Handbuch der Zoologie,* Ed. W. Kükenthal and T. Krumbach, **3** (2): 263–502. (G)
11. * Duncan, H., 1957. *Mem. geol. Soc. Am.,* **67** (2): 783–800. (Palaeoecology)

12. Ellis, J., 1755. *An essay towards a natural history of the corallines.*
13. Franzén, Å., 1956. *Zool. Bidr. Upps.*, **31**: 356–482. (Spermiogenesis)
14. Harmer, S. F., 1896. In *The Cambridge Natural History*, Ed. S. F. Harmer and A. E. Shipley, **2**: 463–533.
15. Harmer, S. F., 1930. *Proc. Linn. Soc. Lond.*, **141**: 68–118. (Review)
16. Harmer, S. F., 1931. *Proc. Linn. Soc. Lond.*, **143**: 113–168. (Review)
17. * Hyman, L. H., 1959. *The invertebrates*, **5**, *Smaller coelomate groups.*
18. Kummel, B., and Raup, D., 1965. *Handbook of paleontological techniques.*
19. Larwood, G. P., Medd, A. W., Owen, D. E., and Tavener-Smith, R., 1967. In *The fossil record*, Ed. W. B. Harland *et al.*: 379–395. (Stratigraphical range of bryozoan families)
20. Mayr, E., 1968. *Syst. Zool.*, **17**: 213–216. (Bryozoa should be name of the phylum)
21. Moore, R. C., Lalicker, C. G., and Fisher, A. G., 1952. *Invertebrate fossils.*
22. Rhodes, F. H. T., 1967. In *The fossil record*, Ed. W. B. Harland *et al.*: 57–76. (Permo-Triassic extinction)
23. * Ryland, J. S., 1967. *Oceanogr. mar. Biol., Ann. Rev.*, **5**: 343–369. (Review)
24. Shrock, R. R., and Twenhofel, W. H., 1953. *Principles of invertebrate paleontology.*
25. Thompson, J. V., 1830 (facsimile 1968). *On Polyzoa . . . Zoological researches*, **5**: 89–102.

GYMNOLAEMATA (EXCEPT FOSSIL)

26. Atkins, D., 1932. *Quart. Jl microsc. Sci.*, **75**: 393–423. (Ciliary feeding)
27. Atkins, D., 1955. *J. mar. biol. Ass., U.K.*, **34**: 441–449. (Feeding of cyphonautes)
28. Banta, W. C., 1968. *J. Morph.*, **125**: 497–508. (Body wall)
28A. Banta, W. C., 1969. *J. Morph.*, **129**: 149–170. (Interzooidal pores)
29. Bobin, G., and Prenant, M., 1952. *Archs Zool. exp. gén.*, **89**: 175–202. (Gizzard of *Bowerbankia*. F)
30. Bobin, G., and Prenant, M., 1968. *Archs Zool. exp. gén.*, **109**: 157–191. (Body wall. F)
31. Braem, F., 1951. *Zoologica, Stuttg.*, **102**: 1–59. (Brackish water Bryozoa. G)
32. Brattström, H., 1954. *Acta. Univ. lund.*, N.S., **50** (9): 1–29. (*Victorella*)
33. Bronstein, G., 1938. *C. r. hebd. Séanc. Acad. Sci., Paris*, **209**: 602–603. (Physiological gradients. F)
34. Bullivant, J. S., 1967. *Ophelia*, **4**: 139–142. (Sperm release)
35. Bullivant, J. S., 1968. *N.Z. Jl mar. freshwat. Res.*, **2**: 111–134. (Feeding in *Zoobotryon*)
36. Calvet, L., 1900. *Contribution à l'histoire naturelle des Bryozoaires Ectoproctes marins.* (F)

7. Carrada, C. C., and Sacchi, C. F., 1964. *Vie Milieu,* **15**: 389–426. (Ecology of *Victorella.* F)

8. Cook, P. L., 1963. *Cah. Biol. mar.,* **4**: 407–413. (Ecology of lunulitiform Bryozoa)

9. Cook, P. L., 1964. *Bull. Br. Mus. nat. Hist.* (Zool.), **12**: 1–35. (Bryozoa and pagurids)

0. Cook, P. L., 1968. *Atlantide Rep.,* **10**: 115–262. (Bryozoa from tropical West Africa)

1. Crisp, D. J., and Williams, G. B., 1960. *Nature, Lond.,* **182**: 1206–1207. (Seaweed extract promotes larval settlement)

2. Eggleston, D., 1963. *The marine Polyzoa of the Isle of Man.* Unpublished thesis, University of Liverpool.

3. Franzén, Å., 1960. *Zool. Bidr. Upps.,* **33**: 135–147. (*Monobryozoon*)

4. Fry, W. G., 1965. *Bull. Br. Mus. nat. Hist.* (Zool.), **12**: 195–223. (Pycnogonids feeding on bryozoans)

5. * Gautier, Y. V., 1962. *Recl Trav. Stn mar. Endoume,* **32**: 1–434. (Ecology and systematics of Mediterranean Bryozoa. F)

6. Gordon, D. P., 1968. *Nature, Lond.,* **219**: 633–634. (Zooidal dimorphism)

7. Gordon, D. P., 1968. *Growth, regeneration and population biology of cheilostomatous polyzoans.* Unpublished thesis, University of Auckland.

8. Greeley, R., 1967. *Bull. geol. Soc. Am.,* **78**: 1179–1182. (Natural orientation of lunulitiform bryozoans)

9. Greeley, R., 1970. *J. Paleont.,* **44**: 343–345. (Natural orientation of lunulitiform bryozoans)

0. Harmer, S. F., 1902. *Quart. Jl microsc. Sci.,* **46**: 263–350. (Cheilostome morphology)

1. Harmer, S. F., 1957. *Siboga Exped.,* **28**: 641–1147. (Indo-Malaysian Ascophora and Gymnocystidea)

2. Hastings, A. B., 1927. *Trans. zool. Soc., Lond.,* **20**: 331–354. (Suez Canal Bryozoa)

3. Hastings, A. B., 1930. *Ann. Mag. nat. Hist.,* Ser. 10, **5**: 552–560. (Hydroid *Zanclea* commensal with bryozoans)

4. Hastings, A. B., 1943. *'Discovery' Rep.,* **22**: 301–510. (Antarctic Cellularioidea)

5. Hastings, A. B., 1963. *Ann. Mag. nat. Hist.,* Ser. 13, **6**: 177–184. (Setiform heterozooids)

6. Lagaaij, R., 1963. *Palaeontology,* **6**: 172–217. (*Cupuladria*)

7. Lagaaij, R., and Gautier, Y. V., 1965. *Micropaleontology,* **11**: 39–58. (Bryozoa from off the Rhône delta)

8. Levinsen, G. M. R., 1909. *Morphological and systematic studies on the cheilostomatous Bryozoa.*

9. Lutaud, G., 1953. *Archs Zool. exp. gén.,* **91**: 36–50. (Calcification in *Escharoides.* F)

0. Lutaud, G., 1961. *Annls Soc. r. zool. Belg.,* **91**: 157–300. (Growth in *Membranipora.* F)

60A. Lutaud, G., 1969. *Z. Zellforsch.*, **99**: 302–314. (Nervous system)
61. Lynch, W. F., 1947. *Biol. Bull. mar. biol. Lab., Woods Hole,* **92**: 115 150. (Metamorphosis in *Bugula*)
62. Mangum, C. P., and Schopf, T. J. M., 1967. *Nature, Lond.,* **213**: 264 (Respiration)
63. Marcus, E., 1926. *Zool. Jb.* (Syst.), **52**: 1–102. (Physiological studie. G)
64. Marcus, E., 1939. *Bolm Fac. Filos. Ciênc. Univ. S Paulo* (Zool.), 3 111–299. (Systematics and general biology. In Portuguese)
65. Marcus, E., and Marcus, du B.-R., 1962. *Bolm Fac. Filos. Ciênc Univ. S Paulo* (Zool.), **24**: 281–324. (Lunulitiform Bryozoa)
66. Massaro, T. A., and Fat, I., 1967. *Nature, Lond.,* **216**: 59. (Respiration
67. Menon, N. R., and Nair, N. B., 1967. *Int. Revue ges. Hydrobio. Hydrogr.,* **52**: 237–256. (Ecology of *Victorella*)
68. Osburn, R. C., 1957. *Mem. geol. Soc. Am.,* **67** (1): 1109–1111. (Distri bution)
69. Powell, N. A., 1967. '*Discovery*' *Rep.,* **34**: 199–394. (New Zealan Ascophora)
70. Powell, N. A., and Cook, P. L., 1966. *Cah. Biol. mar.,* **7**: 53–5 (Polymorphism in *Thalamoporella*)
70A. Rucker, J. B., and Carver, R. E., 1969. *J. Paleont.,* **43**: 791–79 (Carbonate mineralogy)
71. Ryland, J. S., 1958. *Ann. Mag. nat. Hist.,* Ser. 13, **1**: 552–556. (Colou of embryos)
72. Ryland, J. S., 1959. *J. exp. Biol.,* **36**: 613–631. (Larval settlement o algae)
73. Ryland, J. S., 1960. *J. exp. Biol.,* **37**: 783–800. (Light and larva behaviour)
74. Ryland, J. S., 1962. *J. anim. Ecol.,* **31**: 331–338. (Bryozoa an algae)
75. Ryland, J. S., 1963. *Sarsia,* **14**: 1–59. (Ecology and systematics o Norwegian Bryozoa)
76. Ryland, J. S., 1965. *Catalogue of main marine fouling organisms,* 2 (Fouling Bryozoa)
77. Ryland, J. S., 1967. *Nature, Lond.,* **216**: 1040–1041. (Respiration)
78. Ryland, J. S., 1970. In *Marine borers, fungi and fouling organism.* Ed. E. B. Gareth Jones and S. K. Eltringham. (Fouling Bryozoa)
79. Ryland, J. S., and Stebbing, A. R. D., 1970. *Proc. 4th European mar Biol. Symp.* (Oriented growth)
80. Schneider, D., 1963. In *The Lower Metazoa,* Ed. E. C. Dougherty 357–371. (Growth in *Bugula*)
81. Schopf, T. J. M., 1969. *J. Paleont.,* **43**: 234–244. (Ecology)
82. * Schopf, T. J. M., and Manheim, F. T., 1967. *J. Paleont.,* **41** 1197–1225. (Composition of body wall)
83. Schopf, T. J. M., and Travis, D. F., 1968. *Biol. Bull. mar. biol. Lab Woods Hole,* **135**: 436. (Frontal wall in *Schizoporella*)
84. Silén, L., 1938. *Zool. Bidr. Upps.,* **17**: 149–366. (Avicularia. G)

85. Silén, L., 1942. *Zool. Bidr. Upps.*, **22**: 1–59. (Cheilostome-Ctenostome relationships)
86. Silén, L., 1944. *K. svenska VetenskAkad. Handl.*, **21** (6): 1–111. (*Labiostomella*)
87. Silén, L., 1944. *Ark. Zool.*, **35A** (12): 1–40. (Alimentary canal)
88. Silén, L., 1944. *Ark. Zool.*, **35A** (17): 1–34. (Ooecia)
89. Silén, L., 1944. *Zool. Bidr. Upps.*, **22**: 433–488. (Interzooidal connexions)
90. Silén, L., 1947. *Ark. Zool.*, **40A** (4): 1–48. (Shell-boring bryozoans)
91. Silén, L., 1966. *Ophelia*, **3**: 113–140. (Fertilization)
92. Soule, J. D., 1954. *Bull. Sth. Calif. Acad. Sci.*, **53**: 13–34. (Development and classification of the Ctenostomata)
93. Soule, J. D., 1969. *Bull. Sth. Calif. Acad. Sci.*, **67**: 178–181. (*Terebripora*)
94. Stach, L., 1936. *J. Geol.*, **44**: 60–65. (Colony form and habitat)
95. Teal, J. M., 1967. *Nature, Lond.*, **216**: 1239. (Respiration)

GYMNOLAEMATA (FOSSIL)

96. Berthelsen, O., 1962. *Danm. geol. Unders.*, Ser. 2, **83**: 1–290. (Danian Cheilostomata)
97. Brown, D. A., 1952. *The Tertiary cheilostomatous Polyzoa of New Zealand.*
98. Buge, É., 1957. *Mém. Mus. natn Hist. nat. Paris*, Ser. C, **6**: 1–435. (Miocene Bryozoa from north-west France. F)
99. Busk, G., 1859. *A monograph of the fossil Polyzoa of the Crag.*
100. Canu, F., and Bassler, R. S., 1920. *Bull. U.S. natn Mus.*, **106**: 1–879. (North American Tertiary Bryozoa)
101. Cheetham, A. H., 1963. *Mem. geol. Soc. Am.*, **91**: 1–113. (Eocene Bryozoa from Gulf Coast of America)
102. Cheetham, A. H., 1966. *Bull. Br. Mus. nat. Hist.* (Geol.), **13**: 115. (Eocene Cheilostomata from Sussex)
103. Cheetham, A. H., 1968. *Smithson. misc. Collns*, **153** (1): 1–121. (*Metrarabdotos*)
104. Condra, G. E., and Elias, M. K., 1944. *Bull. geol. Soc. Am.*, **55**: 517–568. (Permian and Carboniferous Ctenostomata)
105. David, L., 1965. *Trav. Lab. Géol. Univ. Lyon*, **12**: 33–86. (Miocene Bryozoa from the Rhône valley. F)
106. Greeley, R., 1969. *J. Paleont.*, **43**: 252–256. (Calcification in lunulitiform Bryozoa)
107. Lagaaij, R., 1952. *Meded. geol. Sticht.*, Ser. C, **5** (5): 1–233. (Pliocene Bryozoa of the Low Countries)
108. Lang, W. D., 1921–1922. *Catalogue of the fossil Bryozoa in the British Museum (Natural History). The Cretaceous Bryozoa*, **3** and **4**. (Cribrimorpha)
109. Larwood, G. P., 1962. *Bull. Br. Mus. nat. Hist.* (Geol.), **6**: 1–285. (Cretaceous Cribrimorpha)

110. Thomas, H. D., and Larwood, G. P., 1960. *Palaeontology*, **3**: 370–386. (*Pyripora* and *Rhammatopora*)
111. Vigneaux, M., 1949. *Mém. Soc. géol. Fr.*, **60**: 1–155. (Miocene Bryozoa from the Bassin d'Aquitaine. F)
112. Voigt, E., 1930. *Leopoldina*, **6**: 379–579. (Upper Cretaceous Cheilostomata. G)
113. Voigt, E., 1966. *Neues Jb. Geol. Paläont. Abh.*, **125**: 401–422. (Preservation of Cretaceous Ctenostomata by immuration. G)
114. Voigt, E., 1968. *Bull. Br. Mus. nat. Hist.* (Geol.), **17**: 1–45. (Cretaceous age of supposedly Jurassic Cheilostomata)
115. Voigt, E., 1968. *Neues Jb. Geol. Paläont. Abh.*, **132**: 87–96. (*Arachnidium* from the Lower Cretaceous. G)

STENOLAEMATA

116. Astrova, G. G., 1966. *Int. Geol. Rev.*, **7**: 1622–1628. (Cystoporata)
117. Astrova, G. G., and Morozova, I. P., 1956. *Dokl. Akad. Nauk SSSR*, **110**: 661–664. (Classification of the Cryptostomata. In Russian
118. Boardman, R. S., 1960. *Prof. Pap. U.S. geol. Surv.*, **340**: 1–87 (Trepostomata)
119. Borg, F., 1926. *Zool. Bidr. Upps.*, **10**: 181–507. (Morphology of the Cyclostomata)
120. Borg, F., 1932. *Ark. Zool.*, **24B** (3): 1–6. (*Heteropora*)
121. * Borg, F., 1944. *Further zool. Results Swed. Antarct. Exped.*, **3** (5) 1–276. (Recent Cyclostomata)
122. Borg, F., 1965. *Ark. Zool.*, Ser. 2, **17** (1): 1–91. (Astogenetic studies on Cystoporata and Trepostomata)
123. Cuffey, R. J., 1967. *Paleont. Contr. Univ. Kans.*, Bryozoa, **1**. (*Tabulipora*)
124. Cumings, E. R., 1904. *Am. J. Sci.* (4) **17**: 49–78. (Fenestelloid astogeny)
125. Cumings, E. R., 1905. *Am. J. Sci.* (4) **20**: 169–177. (Fenestelloid astogeny)
126. Cumings, E. R., 1912. *Bull. geol. Soc. Am.*, **23**: 357–370. (Trepostome astogeny)
127. Cumings, E. R., and Galloway, J. J., 1915. *Bull. geol. Soc. Am.*, **26** 349–374. (Trepostome morphology)
128. Elias, M. K., and Condra, G. E., 1957. *Mem. geol. Soc. Am.*, **70** 1–158. (*Fenestella*)
129. Gregory, J. W., 1899 and 1909. *Catalogue of the fossil Bryozoa in the British Museum* (*Natural History*). *The Cretaceous Bryozoa*, **1** and **2** (Cyclostomata)
130. Illies, G., 1968. *Oberrhein. geol. Abh.*, **17**: 217–249. (Astogeny in Tubuliporoidea. G)
131. Levinsen, G. M. R., 1912. *K. danske Vidensk. Selsk. Skr.*, Ser. 7 (Naturv. Math. Afd.), **10** (1): 1–51. (*Meliceritites*)
132. Miller, T. G., 1962. *Palaeontology*, **5**: 540–549. (*Archaeofenestella*)

132A. Nielsen, C., 1970. *Ophelia*, **7**. (Metamorphosis and ancestrula in Cyclostomata)
133. Ross, J. R. P., 1964. *Bull. geol. Soc. Am.*, **75**: 927–948. (Early Palaeozoic Stenolaemata)
134. Ross, J. R. P., 1966. *Okla. Geol. Notes*, **26**: 218–224. (Ordovician ceramoporoid from North America)
135. Tavener-Smith, R., 1969. *Palaeontology*, **12**: 281–309. (Structure and growth in *Fenestella*)
136. Tavener-Smith, R., 1969. *Lethaia*, **2**: 89–97. (Acanthopores of *Leioclema*)
136A. Walter, B., 1969. *Docums Lab. Géol. Fac. Sci. Lyon*, **35**: 1–328. (Jurassic Cyclostomata of France. F)

PHYLACTOLAEMATA

137. Allman, G. J., 1856. *A monograph of the fresh-water Polyzoa.*
138. Brien, P., 1953. *Annl Soc. r. zool. Belg.*, **84**: 301–444. (Comprehensive study of Phylactolaemata. F)
139. Brown, C. J. D., 1933. *Trans. Am. microsc. Soc.*, **52**: 271–314. (Ecology and statoblasts)
140. Bushnell, J. H., 1964. *Am. Zool.*, **4**: 255. (Zoogeography)
141. Bushnell, J. H., 1966. *Ecol. Monogr.*, **36**: 95–123. (Ecology of *Plumatella*)
142. Harmer, S. F., 1913. *Proc. zool. Soc. Lond.*, **1913**: 426–457. (Bryozoa in waterworks)
143. * Lacourt, A. W., 1968. *Zool. Verh., Leiden*, **93**: 1–159. (Systematic monograph)
144. Wiebach, F., 1958. In *Die Tierwelt Mitteleuropas*, Ed. P. Brohmer, **1** (8): 1–56 (G)

BRYOZOA IN THE BRITISH ISLES

145. Atkins, D., 1955. *J. mar. biol. Ass. U.K.*, **34**: 441–449. (Cyphonautes larvae)
146. Hincks, T., 1880. *A history of the British marine Polyzoa.*
147. Hurrell, H. E., 1927. *Jl R. microsc. Soc.*, **47**: 135–142. (Freshwater Bryozoa in East Anglia)
148. Johnston, G., 1847. *A history of British zoophytes*, 2nd ed.
149. Ryland, J. S., 1962. *Fld Stud.*, **1** (4): 33–51. (Identification of intertidal bryozoans)
150. Ryland, J. S., 1964. *J. mar. biol. Ass. U.K.*, **44**: 645–654. (Cyphonautes larvae)
151. Ryland, J. S., 1965. *Fich. Ident. Zooplancton*, **107**: 1–5. (Cyphonautes larvae)
152. * Ryland, J. S., 1969. *Bull. Br. Mus. nat. Hist.* (Zool.), **17**: 207–260. (Supplement to Hincks)[146]

THE ENVIRONMENT

153. Colebrook, J. M., and Robinson, G. A., 1965. *Bull. mar. Ecol.,* **6**: 123–139. (Seasonal cycles of phytoplankton production)
154. Draper, L., 1967. *Marine Geol.,* **5**: 133–140. (Wave activity on the sea bed)
155. Kitching, J. A., and Ebling, F. J., 1967. *Advances in ecological research,* **4**: 197–291. (Ecology of Lough Ine, Ireland)
156. Lagaaij, R., 1968. *Proc. K. ned. Akad. Wet.,* Ser. B, **71** (1): 31–50. (Bryozoan fragments tracing sand movement)
157. Miller, M. C., 1961. *J. anim. Ecol.,* **30**: 95–116. (The food of nudibranchs)
158. Pérès, J. M., 1967. *Oceanogr. mar. Biol., Ann. Rev.,* **5**: 449–533. (Mediterranean benthos)
159. Thorson, G., 1963. In *The undersea challenge,* Ed. B. Eaton: 22–31. (Benthic ecology)
160. Woods Hole Oceanographic Institute, 1952. *Marine fouling and its prevention.*

INDEX

Page-numbers in **bold type** refer to illustrations, those in *italic* refer to the Appendix (Classification). One page-number against a special term refers to a definition or explanation.